Many processes in materials science and engineering, such as the load deformation behavior of certain solids and structures, exhibit nonlinear characteristics. The computer simulation of such processes therefore requires a deep understanding of both the theoretical aspects of nonlinearity and the associated computational techniques.

This book provides a complete set of exercises and solutions in the field of theoretical and computational nonlinear continuum mechanics, and is the perfect companion to *Nonlinear Continuum Mechanics for Finite Element Analysis*, where the authors set out the theoretical foundations of the subject. It employs notation consistent with the theory book and serves as a great resource to students, researchers, and those in industry interested in gaining confidence by practicing through examples. Instructors of the subject will also find the book indispensable in aiding student learning.

JAVIER BONET is a Professor of Engineering and the Head of the College of Engineering at Swansea University, and a visiting Professor at the Universitat Politecnica de Catalunya in Spain. He has extensive experience teaching topics in structural mechanics, including large strain nonlinear solid mechanics, to undergraduate and graduate engineering students. He has been active in research in the area of computational mechanics for more than 25 years, with contributions in modeling superplastic forming, large strain dynamic analysis, membrane modeling, finite element technology including error estimation, and meshless methods (smooth particle hydrodynamics). He has given invited, keynote, and plenary lectures at numerous international conferences.

ANTONIO J. GIL is a Senior Lecturer in the world-leading Civil and Computational Engineering Centre at Swansea University. To date, he has published 50 papers in various areas of computational mechanics, covering the areas of computational simulation of nanomembranes, biomembranes (heart valves) and superplastic forming of medical prostheses, modeling of smart electro-magneto-mechanical devices, and numerical analysis of fast transient dynamical phenomena. He currently leads an interdisciplinary research group of doctoral and postdoctoral researchers comprising mathematicians, engineers, and computer scientists. Having been awarded the National 1st Prize by the Spanish Ministry of Education in 2000, he has since received a number of other research prizes both as Principal Investigator and Ph.D. supervisor in the field of computational mechanics, including the UK 2011 Philip Leverhulme Prize to young researchers for his contributions in computational modeling.

RICHARD D. WOOD is an Honorary Research Fellow in the Civil and Computational Engineering Centre at Swansea University. He has more than 25 years' experience teaching the course Nonlinear Continuum Mechanics for Finite Element Analysis at Swansea University, which he originally developed at the University of Arizona. The same course is now taught by his co-author, Antonio J. Gil. Dr. Wood's academic career has focused on finite element analysis, Swansea University being one of the original major international centers to develop the subject.

To Catherine, Clare and Doreen

WORKED EXAMPLES IN NONLINEAR CONTINUUM MECHANICS FOR FINITE ELEMENT ANALYSIS

WORKED EXAMPLES IN NONLINEAR CONTINUUM MECHANICS FOR FINITE ELEMENT ANALYSIS

Javier Bonet
Swansea University

Antonio J. Gil
Swansea University

Richard D. Wood
Swansea University

CAMBRIDGE
UNIVERSITY PRESS

CAMBRIDGE
UNIVERSITY PRESS

University Printing House, Cambridge CB2 8BS, United Kingdom

One Liberty Plaza, 20th Floor, New York, NY 10006, USA

477 Williamstown Road, Port Melbourne, VIC 3207, Australia

314-321, 3rd Floor, Plot 3, Splendor Forum, Jasola District Centre, New Delhi - 110025, India

103 Penang Road, #05-06/07, Visioncrest Commercial, Singapore 238467

Cambridge University Press is part of the University of Cambridge.

It furthers the University's mission by disseminating knowledge in the pursuit of education, learning and research at the highest international levels of excellence.

www.cambridge.org
Information on this title: www.cambridge.org/9781107603615

First published 2012

A catalogue record for this publication is available from the British Library

ISBN 978-1-107-60361-5 Paperback

Additional resources for this publication at www.flagshyp.com

CONTENTS

PREFACE

This worked examples text is intended primarily as a companion to the second edition of the textbook *Nonlinear Continuum Mechanics for Finite Element Analysis* by Javier Bonet and Richard D. Wood. However, to be reasonably self-contained, where necessary key equations from the textbook are replicated in each chapter.

Textbook equation numbers given at the beginning of each chapter are indicated in square brackets.

Exercises are presented in a mix of direct (tensor), matrix, or indicial notation, whichever provides the greater clarity. Indicial notation is used only when strictly necessary and with summations clearly indicated.

The textbook is augmented by a website, www.flagshyp.com, which contains corrections, software, and sample input data. Updates to this worked examples text will also be included on the website as necessary.

CHAPTER ONE

INTRODUCTION

In this chapter a number of very simple rigid link-spring structures are considered which illustrate many features of nonlinear behavior often associated with more complex structures.

EXAMPLE 1.1

The structure shown in Figure 1.1 comprises a rigid weightless rod $a - b$ supported by a spring of stiffness k. The force F is positive downward as is the vertical displacement v.

(a) Find the equilibrium equation relating F to the slope angle θ and then plot F against the vertical displacement v.

(b) Also determine the directional derivative of F with respect to a change β in θ.

This simple example illustrates the phenomenon known as *snap-through behavior.*

Solution

(a) When considering a finite deformation problem, the equilibrium equation must be established in the deformed position. Consideration of the vertical equilibrium of joint a gives

$$F = T \sin\theta\ ; \quad T = \frac{F}{\sin\theta}, \tag{1.1}$$

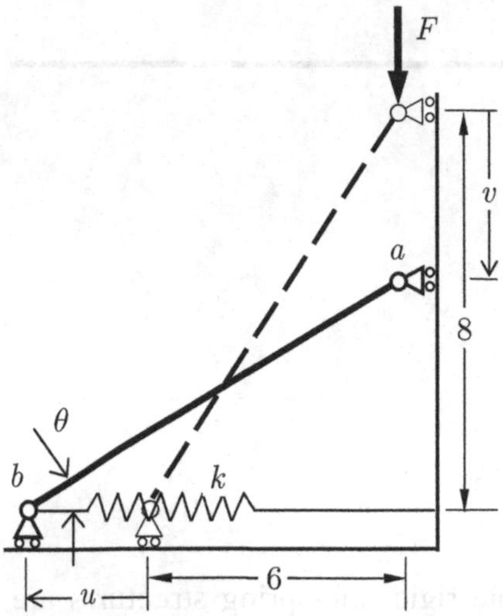

FIGURE 1.1 Rod spring structure

where T is the compressive force in the rod ab. The horizontal equilibrium equation for joint b is found in terms of the tensile force in the spring $S = ku$ as

$$T \cos \theta = S. \quad (1.2)$$

Geometrical considerations yield the displacements u and v in terms of the deformed angle θ as

$$u = 10 \cos \theta - 6\ ;\ v = 8 - 10 \sin \theta. \quad (1.3)$$

Substituting Equation (1.1) into Equation (1.2) gives F as a function of θ as

$$F = k \tan \theta (10 \cos \theta - 6), \quad (1.4)$$

which together with Equation (1.3) enables Figure 1.2 to be drawn.

Note that all points on the plot in Figure 1.2 represent positions of equilibrium. Also observe that for some values of F three equilibrium positions are possible. Indeed, it will be seen in Chapter 3 that multiple positions often are in equilibrium for a given load.

Although not demonstrated here, the positions between the upper and lower peaks are positions of unstable equilibrium. The plot is drawn by

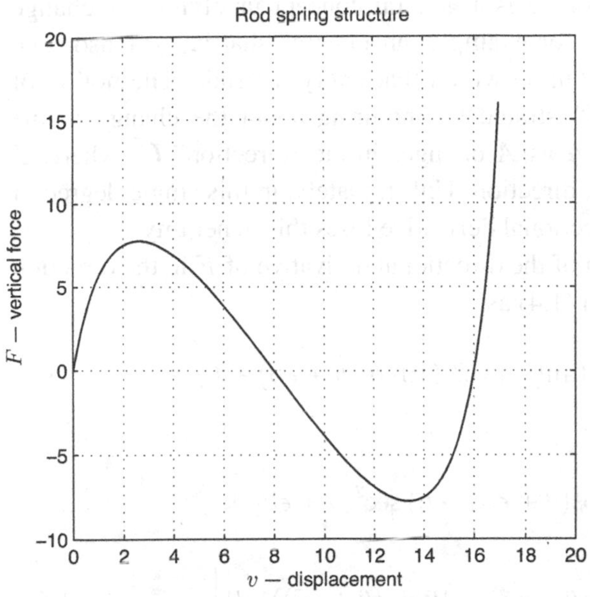

FIGURE 1.2 Rod spring–equilibrium path

choosing a value of θ and calculating F and v; however in practice it is more likely that the force F will determine the value of θ and hence v. The resulting nonlincar solution technique will involve the determination of v as F is gradually incremented at least until the first peak is reached at about $v \approx 2.5$ when a small increase in F will result in a solution in the region of $v \approx 16.5$. This sudden movement (dynamic in reality) is called *snap-through behavior*. Although shallow domes can exhibit such undesirable behavior, snap-through behavior is beneficially employed in everyday life, in light switches, bottle caps, children's hair clips, and many other applications.

(b) The single degree of freedom nature of this example enables nonlinear solutions to be discovered extremely easily, however in reality this is not the case and nonlinear solution techniques need to be devised. Here the Newton–Raphson iterative procedure predominates, which requires the concept of linearization of a nonlinear function which involves the directional derivative. Examples of the directional derivative are presented in some detail in the next chapter, but a brief example is included here.

The directional derivative is a general concept involving the change in a mathematical entity, for example, an integral, matrix, or tensor due to a change in a variable upon which that entity depends. The notion of a "directed" change is illustrated by inquiring about the change of the determinant of a matrix $\boldsymbol{A}$ as $\boldsymbol{A}$ changes in the "direction" $\boldsymbol{U}$, where $\boldsymbol{U}$ is a matter of choice, i.e., direction. Unfortunately, in this single degree of freedom example the directional derivative loses this generality.

The formal definition of the directional derivative of F in the direction β, is given from Equation (1.4) as

$$D(F)[\beta] = \frac{d}{d\epsilon}\bigg|_{\epsilon=0} k\tan(\theta+\epsilon\beta)(10\cos(\theta+\epsilon\beta)-6), \tag{1.5}$$

giving

$$D(F)[\beta] = k\bigg[(10\cos(\theta+\epsilon\beta)-6)\sec^2(\theta+\epsilon\beta)\beta$$

$$+k\left(\tan(\theta+\epsilon\beta)(-10\sin(\theta+\epsilon\beta))\right)\beta\bigg]\bigg|_{\epsilon=0} \tag{1.6a}$$

$$= K(\theta)\beta, \tag{1.6b}$$

$$\text{where } K(\theta) = k\left((10\cos\theta-6)\sec^2\theta - 10\tan\theta\sin\theta\right). \tag{1.6c}$$

Insofar as Equation (1.6b) is the change in F at some position θ due to a change β, then $K(\theta)$ is the stiffness at position θ.

EXAMPLE 1.2

Figure 1.3 shows a weightless rigid column supported by a torsion spring at the base. In the unloaded position the column has an initial imperfection of θ_0. The length of the column is 10, the torsional stiffness 10, and the initial imperfection is $\theta_0 = 0.01$ rads. This example is a simple model of the nonlinear behavior of a vertical column under the action of an axial load.

(a) Find the rotational equilibrium equation and plot the force P against the lateral displacement u.

(b) Linearize the equilibrium equation and set out in outline a Newton–Raphson procedure to solve the equilibrium equation.

(c) Write a computer program to implement the Newton–Raphson solution.

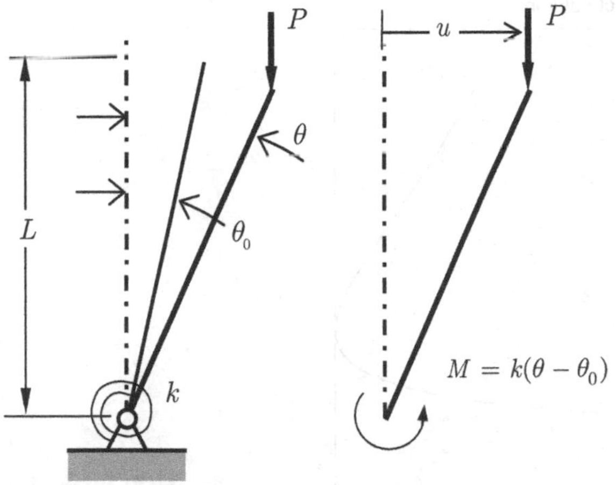

FIGURE 1.3 Imperfect column

Solution

(a) The equilibrium equation is easily found by taking moments about the base to give

$$k(\theta - \theta_0) = PL\sin\theta. \tag{1.7}$$

To plot P against u for the given values of k, L, and θ_0, simply choose a value of θ and solve for P to give

$$P = \frac{k}{L\sin\theta}(\theta - \theta_0)\ ;\ u = L\sin\theta. \tag{1.8}$$

This is shown in Figure 1.4. For a perfect column $\theta_0 = 0$ and for small angles θ the equilibrium equation approximates as $(k - PL)\theta = 0$. Since $\theta \neq 0$, then $P = k/L$, which for this structure is the classical buckling load $P_{\text{critical}} = 1$. Observe that the exact nonlinear solution clearly shows that in the region of P_{critical} a small increase in load produces a large increase in deflection.

(b) In order to conform with the nomenclature used in the program given below, the internal (resisting) moment and external (applied) moment are written as

$$T(\theta) = k(\theta - \theta_0)\ ;\ F = PL\sin\theta. \tag{1.9}$$

This enables the equilibrium Equation (1.7) to be rewritten in terms of a residual moment $R(\theta)$ suitable for the development of the Newton–Raphson

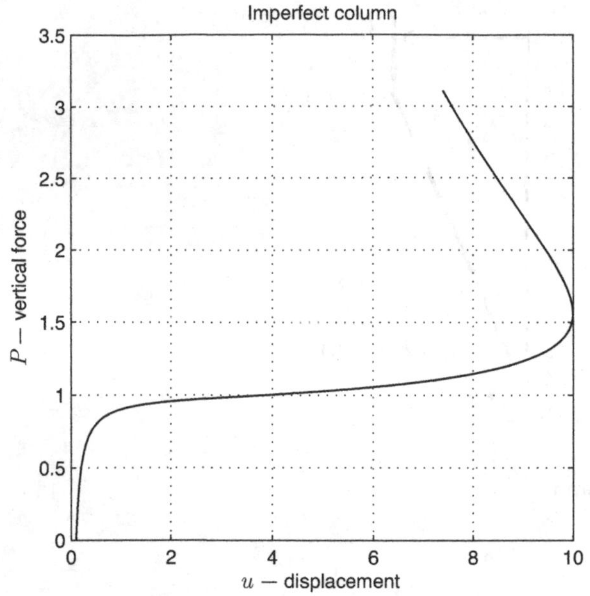

FIGURE 1.4 Imperfect column–equilibrium path

procedure as

$$R(\theta) = T(\theta) - F = 0. \tag{1.10}$$

For a given applied force P, the residual moment $R(\theta)$ is the error in the equilibrium equation due to an incorrect choice of θ. The Newton–Raphson procedure seeks to systematically correct this incorrect choice in order to satisfy the equilibrium equation. To this end, Equation (1.10) is linearized as follows:

$$R(\theta + \Delta\theta) \approx R(\theta) + K(\theta)\Delta\theta = 0. \tag{1.11}$$

where $K(\theta) = DR(\theta)[\Delta\theta]$ is the directional derivative of $R(\theta)$ in the direction $\Delta\theta$ which is found as

$$DR(\theta)[\Delta\theta] = \frac{d}{d\epsilon}\bigg|_{\epsilon=0} [k(\theta + \epsilon\Delta\theta - \theta_0) - PL\sin(\theta + \epsilon\Delta\theta)] \tag{1.12a}$$

$$= [k\Delta\theta - PL\cos(\theta + \epsilon\Delta\theta)\Delta\theta]\bigg|_{\epsilon=0} \tag{1.12b}$$

$$= (k - PL\cos\theta)\Delta\theta = K(\theta)\Delta\theta. \tag{1.12c}$$

A schematic nonlinear solution can now be established using the Newton–Raphson procedure. The force P is applied in a series of increments in order to trace out the complete equilibrium path. The Newton–Raphson iteration is contained within the DO WHILE loop.*

BOX 1.1: Newton-Raphson Algorithm for column problem

- INPUT L, k, θ_0, ΔP and solution parameters (tolerance)
- INITIALIZE $P = 0$, $\theta = \theta_0$ (initial geometry), $R(\theta) = 0$
- FIND initial $K(\theta)$
- LOOP over load increments
    - SET $P = P + \Delta P$
    - SET $R = T(\theta) - F$
    - DO WHILE ($\|R(\theta)\|/\|F\| >$ tolerance)
        - SOLVE $K(\theta)\Delta\theta = -R(\theta)$
        - UPDATE $\theta = \theta + \Delta\theta$
        - FIND $T(\theta)$ and $K(\theta)$
        - FIND $R(\theta) = T(\theta) - F$
    - ENDDO
- ENDLOOP

(c) The schematic nonlinear solution procedure given above is expanded into the FORTRAN program that follows. The nomenclature is largely self-explanatory and follows the symbols used in Equations (1.9) to (1.12c) with the exception that `stiff`=K.

Computer Program for column problem

```
      program Newton Raphson
c     NR program for eccentric simple column
      implicit real*8(a-h,o-z)
      open(10,file='column.out',status='unknown',form='formatted')
c     data c
      tolerance=1.0e-06
      spring=10.0
```

* In Box 1.1 in the textbook, any remaining residual $R(\theta)$ less than the tolerance is carried over into the next load increment.

```
      slength=10.0
      theta0=0.01
      finc=0.02
      ninc=200
      miter=20
c     initialization
      force=0.0
      residual=0.0
      theta=theta0
      stiff=spring-force*slength*cos(theta)
      write(10,'(i5,4f10.5,i5)')0,
     &         theta,force,slength*sin(theta),force,0
c     load loop
      do incrm=1,ninc
         force=force+finc
         t_internal=spring*(theta-theta0)
         f_external=force*slength*sin(theta)
         residual=t_internal-f_external
c     N-R iteration loop
         niter=0
         do while((abs(residual).gt.tolerance).and.(niter.lt.miter))
            niter=niter+1
            dtheta=-residual/stiff
            theta=theta+dtheta
            t_internal=spring*(theta-theta0)
            f_external=force*slength*sin(theta)
            residual=t_internal-f_external
            stiff=spring-force*slength*cos(theta)
         enddo
         if(niter.ge.miter)then
           print *, incrm,' no convergence'
         endif
         write(10,'(i5,4f10.5,i5)')incrm,
     &         theta,force,slength*sin(theta),force,niter
      enddo
      close (10)
      stop
      end
```

CHAPTER TWO

MATHEMATICAL PRELIMINARIES

This chapter presents worked solutions to problems involving vector and tensor algebra, linearization, and the concept of the directional derivative and basic tensor analysis expressions.

Equation summary

Scalar (dot) product [2.5]

$$\begin{aligned}\boldsymbol{u} \cdot \boldsymbol{v} &= \left(\sum_{i=1}^{3} u_i \, \boldsymbol{e}_i \right) \cdot \left(\sum_{j=1}^{3} v_j \, \boldsymbol{e}_j \right) \\ &= \sum_{i,j=1}^{3} u_i v_j \, (\boldsymbol{e}_i \cdot \boldsymbol{e}_j) \\ &= \sum_{i=1}^{3} u_i v_i = \boldsymbol{v} \cdot \boldsymbol{u}. \end{aligned} \tag{2.1}$$

Transformation of Cartesian axes [2.10, 11a]

$$Q_{ij} = \boldsymbol{e}_i \cdot \boldsymbol{e}'_j, \tag{2.2a}$$

$$\boldsymbol{e}'_j = \sum_{i=1}^{3} (\boldsymbol{e}'_j \cdot \boldsymbol{e}_i) \, \boldsymbol{e}_i = \sum_{i=1}^{3} Q_{ij} \, \boldsymbol{e}_i. \tag{2.2b}$$

Identity tensor [2.30a,b]

$$\boldsymbol{I} = \sum_{i=1}^{3} \boldsymbol{e}_i \otimes \boldsymbol{e}_i \quad \text{or} \quad \boldsymbol{I} = \sum_{i,j=1}^{3} \delta_{ij}\, \boldsymbol{e}_i \otimes \boldsymbol{e}_j. \tag{2.3}$$

Tensor product components [Example 2.3]

$$\begin{aligned}(\boldsymbol{u} \otimes \boldsymbol{v}) &= \left(\sum_{i=1}^{3} u_i\, \boldsymbol{e}_i\right) \otimes \left(\sum_{j=1}^{3} v_j\, \boldsymbol{e}_j\right) \\ &= \sum_{i,j=1}^{3} u_i v_j\, \boldsymbol{e}_i \otimes \boldsymbol{e}_j. \end{aligned} \tag{2.4}$$

Alternative Cartesian basis for second-order tensor [2.41]

$$\boldsymbol{S} = \sum_{i,j=1}^{3} S_{ij}\, \boldsymbol{e}_i \otimes \boldsymbol{e}_j = \sum_{i,j=1}^{3} S'_{ij}\, \boldsymbol{e}'_i \otimes \boldsymbol{e}'_j. \tag{2.5}$$

Alternative tensor components [2.42]

$$[\boldsymbol{S}]' = [\boldsymbol{Q}]^T [\boldsymbol{S}][\boldsymbol{Q}] \quad \text{or} \quad S'_{ij} = \sum_{k,l=1}^{3} Q_{ki} S_{kl} Q_{lj}. \tag{2.6}$$

Properties of the double product and trace of two tensors [2.51]

$$\boldsymbol{A} : \boldsymbol{B} = \text{tr}(\boldsymbol{A}^T \boldsymbol{B}) = \text{tr}(\boldsymbol{B}\boldsymbol{A}^T) = \text{tr}(\boldsymbol{B}^T \boldsymbol{A}) = \text{tr}(\boldsymbol{A}\boldsymbol{B}^T) \tag{2.7a}$$

$$= \sum_{i,\,j=1}^{3} A_{ij} B_{ij}. \tag{2.7b}$$

Symmetric second-order tensor in principal directions [2.59]

$$\boldsymbol{S} = \sum_{i=1}^{3} \lambda_i\, \boldsymbol{n}_i \otimes \boldsymbol{n}_i. \tag{2.8}$$

Cartesian basis for third-order tensor [2.64]

$$\boldsymbol{\mathcal{A}} = \sum_{i,j,k=1}^{3} \mathcal{A}_{ijk}\, \boldsymbol{e}_i \otimes \boldsymbol{e}_j \otimes \boldsymbol{e}_k. \tag{2.9}$$

Directional derivative of a general nonlinear function $\mathcal{F}(\mathbf{x})$ at $\mathbf{x}_0$ in the direction of $\mathbf{u}$ [2.101]

$$D\mathcal{F}(\mathbf{x}_0)[\mathbf{u}] = \frac{d}{d\epsilon}\bigg|_{\epsilon=0} \mathcal{F}(\mathbf{x}_0 + \epsilon\mathbf{u}). \tag{2.10}$$

Gradient of a scalar [2.128]

$$\nabla f = \frac{\partial f}{\partial \boldsymbol{x}}. \tag{2.11}$$

Gradient of a vector [2.130]

$$\nabla \boldsymbol{v} = \sum_{i,j=1}^{3} \frac{\partial v_i}{\partial x_j} \boldsymbol{e}_i \otimes \boldsymbol{e}_j; \quad \nabla \boldsymbol{v} = \frac{\partial \boldsymbol{v}}{\partial \boldsymbol{x}}. \tag{2.12}$$

Divergence of a vector [2.131]

$$\operatorname{div} \boldsymbol{v} = \operatorname{tr} \nabla \boldsymbol{v} = \nabla \boldsymbol{v} : \boldsymbol{I} = \sum_{i=1}^{3} \frac{\partial v_i}{\partial x_i}. \tag{2.13}$$

Gauss or divergence theorem for a vector field $\boldsymbol{v}$ [2.138]

$$\int_V \operatorname{div} \boldsymbol{v}\, dV = \int_{\partial V} \boldsymbol{v} \cdot \boldsymbol{n}\, dA. \tag{2.14}$$

EXAMPLE 2.1: Textbook Exercise 2.1

The second-order tensor $\boldsymbol{P}$ maps any vector $\boldsymbol{u}$ to its projection on a plane passing through the origin and with unit normal $\boldsymbol{a}$. Show that:

$$P_{ij} = \delta_{ij} - a_i a_j; \quad \boldsymbol{P} = \boldsymbol{I} - \boldsymbol{a} \otimes \boldsymbol{a}.$$

Show that the invariants of $\boldsymbol{P}$ are $I_{\boldsymbol{P}} = II_{\boldsymbol{P}} = 2$, $III_{\boldsymbol{P}} = 0$, and find the eigenvalues and eigenvectors of $\boldsymbol{P}$.

Solution

Let the unit normal to the plane Π be $\boldsymbol{a}$ and $\boldsymbol{u}_{\text{proj}}$ be the projection of the vector $\boldsymbol{u}$ on this plane. Consequently,

$$\boldsymbol{u}_{\text{proj}} = \boldsymbol{u} - (\boldsymbol{u} \cdot \boldsymbol{a})\boldsymbol{a}, \tag{2.15}$$

which can be rewritten as

$$\begin{aligned} \boldsymbol{u}_{\text{proj}} &= \boldsymbol{I}\boldsymbol{u} - (\boldsymbol{a} \otimes \boldsymbol{a})\boldsymbol{u} \\ &= (\boldsymbol{I} - \boldsymbol{a} \otimes \boldsymbol{a})\boldsymbol{u}. \end{aligned} \tag{2.16}$$

By defining $\boldsymbol{P} := \boldsymbol{I} - \boldsymbol{a} \otimes \boldsymbol{a}$ it can be seen that the second-order tensor $\boldsymbol{P}$ projects the vector $\boldsymbol{u}$ onto the plane Π as

$$\begin{aligned} \boldsymbol{P}\boldsymbol{u} &= (\boldsymbol{I} - \boldsymbol{a} \otimes \boldsymbol{a})\boldsymbol{u} \\ &= \boldsymbol{u}_{\text{proj}}, \end{aligned} \tag{2.17}$$

which in index notation becomes

$$\begin{aligned} (u_{\text{proj}})_i &= P_{ij} u_j \\ &= u_i - u_j a_j a_i, \end{aligned} \tag{2.18}$$

where $P_{ij} = (\delta_{ij} - a_i a_j)$. To consider the invariants of the second-order tensor $\boldsymbol{P}$, define a triad of orthonormal vectors as $\{\boldsymbol{a}, \boldsymbol{b}, \boldsymbol{c}\}$ where $\boldsymbol{a}$ is the unit normal to the plane Π as given above. Employing Equation (2.3a,b) the identity tensor can be written as

$$\boldsymbol{I} = \boldsymbol{a} \otimes \boldsymbol{a} + \boldsymbol{b} \otimes \boldsymbol{b} + \boldsymbol{c} \otimes \boldsymbol{c}, \tag{2.19}$$

which enables the projection tensor $\boldsymbol{P}$ to be expressed as

$$\begin{aligned} \boldsymbol{P} &= \boldsymbol{I} - \boldsymbol{a} \otimes \boldsymbol{a} \\ &= \boldsymbol{b} \otimes \boldsymbol{b} + \boldsymbol{c} \otimes \boldsymbol{c}. \end{aligned} \tag{2.20}$$

From the above equation $\boldsymbol{P}$ is obviously symmetric. Equation (2.8) implies that $\boldsymbol{P}$ has unique eigenvalues and eigenvectors, and, rewriting Equation (2.20) (pedantically) as

$$\boldsymbol{P} = 1(\boldsymbol{b} \otimes \boldsymbol{b}) + 1(\boldsymbol{c} \otimes \boldsymbol{c}) + 0(\boldsymbol{a} \otimes \boldsymbol{a}), \tag{2.21}$$

it can be seen that $\boldsymbol{P}$ will have three eigenvalues: $\lambda_1 = 1$, $\lambda_2 = 1$, and $\lambda_3 = 0$ together with three corresponding eigenvectors $\boldsymbol{n}_1 = \boldsymbol{b}$ $\boldsymbol{n}_2 = \boldsymbol{c}$, and $\boldsymbol{n}_3$ normal to $\boldsymbol{b}$ and $\boldsymbol{c}$. The invariants can now be enumerated as

$$I_{\boldsymbol{P}} = \text{tr}(\boldsymbol{P}) = \sum_{i=1}^{3} \lambda_i = 2, \tag{2.22a}$$

$$I\!I_{\boldsymbol{P}} = \boldsymbol{P} : \boldsymbol{P} = \sum_{i=1}^{3} \lambda_i^2 = 2, \tag{2.22b}$$

$$I\!I\!I_{\boldsymbol{P}} = \det(\boldsymbol{P}) = \lambda_1 \lambda_2 \lambda_3 = 0. \tag{2.22c}$$

EXAMPLE 2.2: Textbook Exercise 2.2

Using a procedure similar to that employed in Equations (2.5) and (2.6), obtain transformation equations for the components of third- and fourth-order tensors in two sets of bases $\boldsymbol{e}_i$ and $\boldsymbol{e}'_i$ that are related by the 3-D transformation tensor $\boldsymbol{Q}$ with components $Q_{ij} = \boldsymbol{e}_i \cdot \boldsymbol{e}'_j$.

Solution

Define a third-order tensor $\mathcal{G}$ which can be expressed in terms of two different bases $\boldsymbol{e}_i$ and $\boldsymbol{e}'_i$, where $i = 1, 2, 3$. Using Equation (2.9), $\mathcal{G}$ can be expressed in the two bases as

$$\mathcal{G} = \sum_{i,j,k=1}^{3} \mathcal{G}_{ijk}\, \boldsymbol{e}_i \otimes \boldsymbol{e}_j \otimes \boldsymbol{e}_k, \tag{2.23a}$$

$$\mathcal{G} = \sum_{i,j,k=1}^{3} \mathcal{G}'_{ijk}\, \boldsymbol{e}'_i \otimes \boldsymbol{e}'_j \otimes \boldsymbol{e}'_k. \tag{2.23b}$$

Using Equations (2.2) the alternative bases can be related as

$$\boldsymbol{e}'_i = \sum_{j=1}^{3} Q_{ji}\, \boldsymbol{e}_j \quad \text{where} \quad Q_{ij} = \boldsymbol{e}_i \cdot \boldsymbol{e}'_j. \tag{2.24}$$

Substituting the above equation into Equation (2.23) yields

$$\begin{aligned}
\mathcal{G} &= \sum_{i,j,k=1}^{3} \mathcal{G}'_{ijk} \left(\sum_{l=1}^{3} Q_{li}\, \boldsymbol{e}_l \right) \otimes \left(\sum_{m=1}^{3} Q_{mj}\, \boldsymbol{e}_m \right) \otimes \left(\sum_{n=1}^{3} Q_{nk}\, \boldsymbol{e}_n \right) \\
&= \sum_{l,m,n=1}^{3} \left(\sum_{i,j,k=1}^{3} \mathcal{G}'_{ijk}\, Q_{li}\, Q_{mj}\, Q_{nk} \right) \boldsymbol{e}_l \otimes \boldsymbol{e}_m \otimes \boldsymbol{e}_n \\
&= \sum_{l,m,n=1}^{3} \mathcal{G}_{lmn}\, \boldsymbol{e}_l \otimes \boldsymbol{e}_m \otimes \boldsymbol{e}_n,
\end{aligned} \tag{2.25}$$

where

$$\mathcal{G}_{lmn} = \sum_{i,j,k=1}^{3} \mathcal{G}'_{ijk}\, Q_{li}\, Q_{mj}\, Q_{nk}. \tag{2.26}$$

The same expressions can be straightforwardly obtained for a general fourth-order tensor $\mathcal{H}$ as

$$\mathcal{H} = \sum_{i,j,k,l=1}^{3} \mathcal{H}_{ijkl}\, \boldsymbol{e}_i \otimes \boldsymbol{e}_j \otimes \boldsymbol{e}_k \otimes \boldsymbol{e}_l, \tag{2.27a}$$

$$\mathcal{H} = \sum_{i,j,k,l=1}^{3} \mathcal{H}'_{ijkl}\, \boldsymbol{e}'_i \otimes \boldsymbol{e}'_j \otimes \boldsymbol{e}'_k \otimes \boldsymbol{e}'_l, \tag{2.27b}$$

where

$$\mathcal{H}'_{ijkl} = \sum_{m,n,p,q=1}^{3} \mathcal{H}_{mnpq}\, Q_{mi}\, Q_{nj}\, Q_{pk}\, Q_{ql}, \tag{2.28a}$$

$$\mathcal{H}_{ijkl} = \sum_{m,n,p,q=1}^{3} \mathcal{H}'_{mnpq}\, Q_{im}\, Q_{jn}\, Q_{kp}\, Q_{lq}. \tag{2.28b}$$

EXAMPLE 2.3: Textbook Exercise 2.3

If L and l are initial and current lengths, respectively, of an axial rod, the associated Engineering, Logarithmic, Green, and Almansi strains given in Section 1.3.1 are

$$\varepsilon_E(l) = \frac{l-L}{L};\ \varepsilon_L(l) = \ln\frac{l}{L};\ \varepsilon_G(l) = \frac{l^2-L^2}{2L^2};$$
$$\varepsilon_A(l) = \frac{l^2-L^2}{2l^2}.$$

Find the directional derivatives $D\varepsilon_E(l)[u]$, $D\varepsilon_L(l)[u]$, $D\varepsilon_G(l)[u]$, and $D\varepsilon_A(l)[u]$ where u is a small increment in the length l.

Solution

Using Equation (2.10) gives

$$D\varepsilon_E(l)[u] = \frac{d}{d\epsilon}\bigg|_{\epsilon=0} \frac{(l+\epsilon u)-L}{L} = \frac{u}{L}. \tag{2.29a}$$

$$D\varepsilon_L(l)[u] = \frac{d}{d\epsilon}\bigg|_{\epsilon=0} \Big(\ln(l+\epsilon u) - \ln(L)\Big)$$
$$= \left(\frac{u}{l+\epsilon u}\right)\bigg|_{\epsilon=0} = \frac{u}{l}. \tag{2.29b}$$

$$D\varepsilon_G(l)[u] = \frac{d}{d\epsilon}\bigg|_{\epsilon=0} \frac{(l+\epsilon u)^2 - L^2}{2L^2}$$
$$= \frac{2(l+\epsilon u)u}{2L^2}\bigg|_{\epsilon=0}$$
$$= \frac{ul}{L^2} = \left(\frac{l}{L}\right)\left(\frac{u}{l}\right)\left(\frac{l}{L}\right). \tag{2.29c}$$

Observe that this is analogous to textbook Equation (4.72) $D\boldsymbol{E}[\boldsymbol{u}] = \boldsymbol{F}^T \boldsymbol{\varepsilon} \boldsymbol{F}$ where $\boldsymbol{F} = l/L$. Finally,

$$D\varepsilon_A(l)[u] = \frac{d}{d\epsilon}\bigg|_{\epsilon=0} \frac{(l+\epsilon u)^2 - L^2}{2(l+\epsilon u)^2}$$
$$= \frac{d}{d\epsilon}\bigg|_{\epsilon=0} \left(\frac{1}{2} - \frac{L^2}{2}(l+\epsilon u)^{-2}\right)$$
$$= \frac{L^2 u}{(l+\epsilon u)^3}\bigg|_{\epsilon=0}$$
$$= \frac{L^2 u}{l^3} = \left(\frac{L}{l}\right)\left(\frac{u}{l}\right)\left(\frac{L}{l}\right). \tag{2.30}$$

EXAMPLE 2.4: Textbook Exercise 2.4

Given any second-order tensor $\boldsymbol{S}$, linearize the expression $\boldsymbol{S}^2 = \boldsymbol{S}\boldsymbol{S}$ in the direction of an increment $\boldsymbol{U}$.

Solution

$$D\boldsymbol{S}^2[\boldsymbol{U}] = \frac{d}{d\epsilon}\bigg|_{\epsilon=0} (\boldsymbol{S} + \epsilon \boldsymbol{U})(\boldsymbol{S} + \epsilon \boldsymbol{U})$$
$$= \frac{d}{d\epsilon}\bigg|_{\epsilon=0} (\boldsymbol{S}\boldsymbol{S} + \epsilon \boldsymbol{U}\boldsymbol{S} + \epsilon \boldsymbol{S}\boldsymbol{U} + \epsilon^2 \boldsymbol{U}\boldsymbol{U})$$
$$= (\boldsymbol{U}\boldsymbol{S} + \boldsymbol{S}\boldsymbol{U} + 2\epsilon \boldsymbol{U}\boldsymbol{U})\bigg|_{\epsilon=0}$$
$$= \boldsymbol{U}\boldsymbol{S} + \boldsymbol{S}\boldsymbol{U}. \tag{2.31}$$

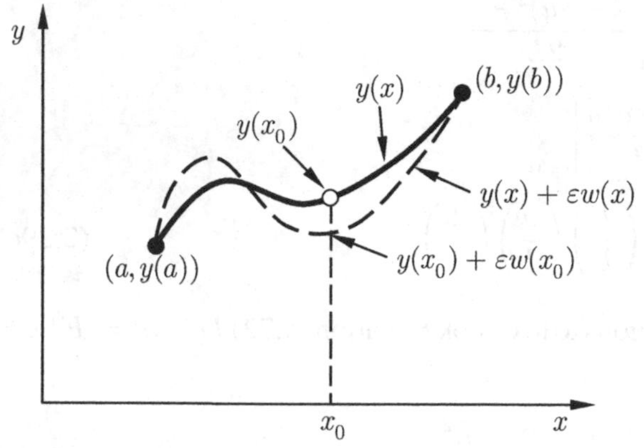

FIGURE 2.1 Example 2.5

EXAMPLE 2.5: Textbook Exercise 2.5

Consider a functional I that when applied to the function $y(x)$ gives the integral:

$$I(y(x)) = \int_a^b f(x, y, y')\, dx,$$

where f is a general expression involving x, $y(x)$ and the derivative $y'(x) = dy/dx$. Show that the function $y(x)$ that renders the above functional stationary and satisfies the boundary conditions $y(a) = y_a$ and $y(b) = y_b$ is the solution of the following Euler–Lagrange differential equation:

$$\frac{d}{dx}\left(\frac{\partial f}{\partial y'}\right) - \frac{\partial f}{\partial y} = 0.$$

Solution

The stationarity of the functional is achieved by making the directional derivative of $I(y(x))$ in the direction of $w(x)$ equal to zero. Here $w(x)$ can be considered as a variation of the function $y(x)$ with $w(a) = w(b) = 0$, see Figure 2.1. Hence

$$DI(y(x))[w(x)] = \left.\frac{d}{d\epsilon}\right|_{\epsilon=0} I(y(x) + \epsilon w(x)) = 0. \tag{2.32}$$

Consequently,

$$\left.\frac{d}{d\epsilon}\right|_{\epsilon=0} I(y(x)+\epsilon w(x)) = \left.\frac{d}{d\epsilon}\right|_{\epsilon=0}\left[\int_a^b f(x,s(x,\epsilon),t(x,\epsilon)dx\right], \tag{2.33}$$

where

$$s(x,\epsilon) = y(x)+\epsilon w(x), \tag{2.34a}$$

$$t(x,\epsilon) = y'(x)+\epsilon w'(x). \tag{2.34b}$$

Hence

$$\left.\frac{d}{d\epsilon}\right|_{\epsilon=0} I(y(x)+\epsilon w(x)) = \int_a^b \left.\frac{\partial}{\partial\epsilon}\right|_{\epsilon=0}\left[f\big(x,s(x,\epsilon),t(x,\epsilon)\big)\right]dx. \tag{2.35}$$

Applying the chain rule to the term inside the integral in Equation (2.35) gives

$$\begin{aligned}\left.\frac{\partial}{\partial\epsilon}\right|_{\epsilon=0}\left[f\big(x,s(x,\epsilon),t(x,\epsilon)\big)dx\right] &= \left.\left(\frac{\partial f}{\partial s}\frac{\partial s}{\partial\epsilon}+\frac{\partial f}{\partial t}\frac{\partial t}{\partial\epsilon}\right)\right|_{\epsilon=0}\\ &= \frac{\partial f}{\partial y}w(x)+\frac{\partial f}{\partial y'}w'(x),\end{aligned} \tag{2.36}$$

enabling Equation (2.33) to be written as

$$\int_a^b \left.\frac{\partial}{\partial\epsilon}\right|_{\epsilon=0}\left[f(x,s(x,\epsilon),t(x,\epsilon)dx\right] = \int_a^b\left[\frac{\partial f}{\partial y}w(x)+\frac{\partial f}{\partial y'}w'(x)\right]dx. \tag{2.37}$$

Integrating the second term in the right-hand side of Equation (2.37) by parts gives

$$\int_a^b \frac{\partial f}{\partial y'}w'(x)dx = \left[\frac{\partial f}{\partial y'}w(x)\right]_a^b - \int_a^b w(x)\frac{d}{dx}\left(\frac{\partial f}{\partial y'}\right)dx. \tag{2.38}$$

Since the boundary conditions $w(a) = w(b) = 0$, the first term above is zero and Equation (2.37) can be rewritten as

$$\begin{aligned}&\int_a^b \left.\frac{\partial}{\partial\epsilon}\right|_{\epsilon=0}\left[f\big(x,s(x,\epsilon),t(x,\epsilon)\big)\right]dx\\ &= \int_a^b w(x)\left[\frac{\partial f}{\partial y}-\frac{d}{dx}\left(\frac{\partial f}{\partial y'}\right)\right]dx\end{aligned} \tag{2.39}$$

Finally the stationarity condition Equation (2.32) can be expressed as

$$DI(y(x))[w(x)] = \int_a^b w(x)\left[\frac{\partial f}{\partial y} - \frac{d}{dx}\left(\frac{\partial f}{\partial y'}\right)\right] dx = 0 \tag{2.40}$$

for any function $w(x)$; consequently,

$$\frac{d}{dx}\left(\frac{\partial f}{\partial y'}\right) - \frac{\partial f}{\partial y} dx = 0. \tag{2.41}$$

EXAMPLE 2.6: Textbook Exercise 2.6

Prove textbook Equations (2.135a–g) following the procedure shown in Example 2.11.

$$\boldsymbol{\nabla}(f\boldsymbol{v}) = f\boldsymbol{\nabla}\boldsymbol{v} + \boldsymbol{v}\otimes\boldsymbol{\nabla}f \tag{2.42a}$$

$$\text{div}\,(f\boldsymbol{v}) = f\text{div}\,\boldsymbol{v} + \boldsymbol{v}\cdot\boldsymbol{\nabla}f \tag{2.42b}$$

$$\boldsymbol{\nabla}(\boldsymbol{v}\cdot\boldsymbol{w}) = (\boldsymbol{\nabla}\boldsymbol{v})^T\boldsymbol{w} + (\boldsymbol{\nabla}\boldsymbol{w})^T\boldsymbol{v} \tag{2.42c}$$

$$\text{div}\,(\boldsymbol{v}\otimes\boldsymbol{w}) = \boldsymbol{v}\,\text{div}\,\boldsymbol{w} + (\boldsymbol{\nabla}\boldsymbol{v})\boldsymbol{w} \tag{2.42d}$$

$$\text{div}\,(\boldsymbol{S}^T\boldsymbol{v}) = \boldsymbol{S} : \boldsymbol{\nabla}\boldsymbol{v} + \boldsymbol{v}\cdot\text{div}\,\boldsymbol{S} \tag{2.42e}$$

$$\text{div}\,(f\boldsymbol{S}) = f\,\text{div}\,\boldsymbol{S} + \boldsymbol{S}\boldsymbol{\nabla}f \tag{2.42f}$$

$$\boldsymbol{\nabla}(f\boldsymbol{S}) = f\boldsymbol{\nabla}\boldsymbol{S} + \boldsymbol{S}\otimes\boldsymbol{\nabla}f. \tag{2.42g}$$

Solution

For Equation (2.42a) using Equation (2.12) express $\boldsymbol{\nabla}(f\boldsymbol{v})$ in indicial notation to give

$$\begin{aligned}
\boldsymbol{\nabla}(f\boldsymbol{v}) &= \sum_{j=1}^{3}\frac{\partial}{\partial x_j}\left(f\sum_{i=1}^{3} v_i\boldsymbol{e}_i\right)\otimes\boldsymbol{e}_j \\
&= \sum_{i,j=1}^{3}\frac{\partial}{\partial x_j}(fv_i)\,\boldsymbol{e}_i\otimes\boldsymbol{e}_j \\
&= \sum_{i,j=1}^{3} f\frac{\partial v_i}{\partial x_j}\,\boldsymbol{e}_i\otimes\boldsymbol{e}_j + \sum_{i,j=1}^{3}\frac{\partial f}{\partial x_j}v_i\,\boldsymbol{e}_i\otimes\boldsymbol{e}_j \\
&= f\sum_{i,j=1}^{3}\frac{\partial v_i}{\partial x_j}\,\boldsymbol{e}_i\otimes\boldsymbol{e}_j + \sum_{i=1}^{3} v_i\,\boldsymbol{e}_i\otimes\sum_{j=1}^{3}\frac{\partial f}{\partial x_j}\boldsymbol{e}_j \\
&= f\boldsymbol{\nabla}\boldsymbol{v} + \boldsymbol{v}\otimes\boldsymbol{\nabla}f.
\end{aligned} \tag{2.43}$$

For Equation (2.42b) using Equation (2.13) gives

$$
\begin{aligned}
\operatorname{div}(f\boldsymbol{v}) &= \sum_{i=1}^{3} \frac{\partial}{\partial x_i}(f v_i) \\
&= \sum_{i=1}^{3} f \frac{\partial v_i}{\partial x_i} + \sum_{i=1}^{3} v_i \frac{\partial f}{\partial x_i} \\
&= f \sum_{i=1}^{3} \frac{\partial v_i}{\partial x_i} + \sum_{i=1}^{3} v_i \frac{\partial f}{\partial x_i} \\
&= f \operatorname{div} \boldsymbol{v} + \boldsymbol{v} \cdot \boldsymbol{\nabla} f. \qquad (2.44)
\end{aligned}
$$

For Equation (2.42c) using Equations (2.1) and (2.11) gives

$$
\begin{aligned}
\boldsymbol{\nabla}(\boldsymbol{v} \cdot \boldsymbol{w}) &= \sum_{i=1}^{3} \frac{\partial}{\partial x_i} \left(\sum_{j=1}^{3} v_j w_j \right) \boldsymbol{e}_i \\
&= \sum_{i,j=1}^{3} \left(\frac{\partial v_j}{\partial x_i} w_j \boldsymbol{e}_i + v_j \frac{\partial w_j}{\partial x_i} \boldsymbol{e}_i \right) \\
&= \sum_{j=1}^{3} \left(\sum_{i=1}^{3} \frac{\partial v_j}{\partial x_i} \boldsymbol{e}_i \right) w_j + \sum_{j=1}^{3} \left(\sum_{i=1}^{3} \frac{\partial w_j}{\partial x_i} \boldsymbol{e}_i \right) v_j \\
&= (\boldsymbol{\nabla} \boldsymbol{v})^T \boldsymbol{w} + (\boldsymbol{\nabla} \boldsymbol{w})^T \boldsymbol{v}. \qquad (2.45)
\end{aligned}
$$

For Equation (2.42d) using textbook Example 2.4 and Equation (2.13) yields

$$
\begin{aligned}
\operatorname{div}(\boldsymbol{v} \otimes \boldsymbol{w}) &= \operatorname{div} \left(\sum_{i=1}^{3} v_i \boldsymbol{e}_i \otimes \sum_{j=1}^{3} w_j \boldsymbol{e}_j \right) \\
&= \sum_{i,j=1}^{3} \frac{\partial}{\partial x_j} (v_i w_j) \boldsymbol{e}_i \\
&= \sum_{i,j=1}^{3} v_i \frac{\partial w_j}{\partial x_j} \boldsymbol{e}_i + \sum_{i,j=1}^{3} \frac{\partial v_i}{\partial x_j} w_j \boldsymbol{e}_i \\
&= \boldsymbol{v} \operatorname{div} \boldsymbol{w} + (\boldsymbol{\nabla} \boldsymbol{v}) \boldsymbol{w}. \qquad (2.46)
\end{aligned}
$$

For Equation (2.42e) see textbook Example 2.11. However, an expanded version is given below:

$$
\begin{aligned}
\operatorname{div}(\boldsymbol{S}^T\boldsymbol{v}) &= \operatorname{div}\left(\left(\sum_{i,j=1}^{3} S_{ji}\,\boldsymbol{e}_i\otimes\boldsymbol{e}_j\right)\sum_{k=1}^{3} v_k\boldsymbol{e}_k\right)\\
&= \operatorname{div}\left(\sum_{i,j,k=1}^{3} S_{ji}\,v_k(\boldsymbol{e}_j\cdot\boldsymbol{e}_k)\,\boldsymbol{e}_i\right)\\
&= \operatorname{div}\left(\sum_{i,j,k=1}^{3} S_{ji}\,v_k\,\delta_{jk}\boldsymbol{e}_i\right)\\
&= \sum_{i,j=1}^{3}\frac{\partial}{\partial x_i}(S_{ji}\,v_j)\\
&= \sum_{i,j=1}^{3} S_{ji}\frac{\partial v_j}{\partial x_i} + \sum_{i,j=1}^{3} v_j\frac{\partial S_{ji}}{\partial x_i}\\
&= \boldsymbol{S}:\nabla\boldsymbol{v} + \boldsymbol{v}\cdot\operatorname{div}(\boldsymbol{S}). \qquad (2.47)
\end{aligned}
$$

For Equation (2.42f),

$$
\begin{aligned}
\operatorname{div}(f\boldsymbol{S}) &= \operatorname{div}\left(f\sum_{i,j=1}^{3} S_{ij}\,\boldsymbol{e}_i\otimes\boldsymbol{e}_j\right)\\
&= \sum_{i,j=1}^{3}\operatorname{div}(fS_{ij}\,\boldsymbol{e}_i\otimes\boldsymbol{e}_j)\\
&= \sum_{i,j=1}^{3}\frac{\partial}{\partial x_j}(fS_{ij}\,\boldsymbol{e}_i)\\
&= \sum_{i,j=1}^{3}\frac{\partial}{\partial x_j}(fS_{ij})\,\boldsymbol{e}_i\\
&= \sum_{i,j=1}^{3}\left(f\frac{\partial S_{ij}}{\partial x_j}\boldsymbol{e}_i + \frac{\partial f}{\partial x_j}S_{ij}\,\boldsymbol{e}_i\right)\\
&= f\sum_{i,j=1}^{3}\frac{\partial S_{ij}}{\partial x_j}\,\boldsymbol{e}_i + \sum_{i,j=1}^{3} S_{ij}\,\boldsymbol{e}_i\frac{\partial f}{\partial x_j}\\
&= f\operatorname{div}(\boldsymbol{S}) + \boldsymbol{S}\,\nabla f. \qquad (2.48)
\end{aligned}
$$

For Equation (2.42g),

$$\begin{aligned}\boldsymbol{\nabla}(f\boldsymbol{S}) &= \sum_{k=1}^{3} \frac{\partial}{\partial x_k}(f\boldsymbol{S}) \otimes \boldsymbol{e}_k \\ &= \sum_{i,j,k=1}^{3} \frac{\partial}{\partial x_k}(fS_{ij})\boldsymbol{e}_i \otimes \boldsymbol{e}_j \otimes \boldsymbol{e}_k \\ &= \sum_{i,j,k=1}^{3} f\frac{\partial S_{ij}}{\partial x_k}\boldsymbol{e}_i \otimes \boldsymbol{e}_j \otimes \boldsymbol{e}_k \\ &\quad + \sum_{i,j=1}^{3} S_{ij}\boldsymbol{e}_i \otimes \boldsymbol{e}_j \otimes \sum_{k=1}^{3} \frac{\partial f}{\partial x_k}\boldsymbol{e}_k \\ &= f\boldsymbol{\nabla S} + \boldsymbol{S} \otimes \boldsymbol{\nabla} f. \end{aligned} \tag{2.49}$$

EXAMPLE 2.7: Textbook Exercise 2.7

Show that the volume of a closed 3-D body V is variously given as

$$V = \int_{\partial V} x\, n_x\, dA = \int_{\partial V} y\, n_y\, dA = \int_{\partial V} z\, n_z\, dA,$$

where n_x, n_y, and n_z are the x, y, and z components of the unit normal $\boldsymbol{n}$.

Solution

The volume of an enclosed body V with boundary ∂V can be obtained as follows:

$$V = \int_V dV. \tag{2.50}$$

Recall the Gauss or divergence theorem for a vector field $\boldsymbol{w}$ given by Equation (2.14) as

$$\int_V \operatorname{div} \boldsymbol{w}\, dV = \int_{\partial V} \boldsymbol{w} \cdot \boldsymbol{n}\, dA \quad \text{where } \boldsymbol{n} = \begin{bmatrix} n_x \\ n_y \\ n_z \end{bmatrix}, \tag{2.51}$$

where the vector $\boldsymbol{n}$ is normal to the surface ∂V of the volume V. Now recall Equation (2.13) for the definition of div $\boldsymbol{w}$:

$$\operatorname{div} \boldsymbol{w} = \sum_{i=1}^{3} \frac{\partial w_i}{\partial x_i}. \tag{2.52}$$

Let the vector $\boldsymbol{w} = [x, 0, 0]^{\mathrm{T}}$; consequently,

$$\int_V \operatorname{div} \boldsymbol{w} \, dV = \int_V dV = V. \tag{2.53}$$

But

$$\int_V \operatorname{div} \boldsymbol{w} \, dV = \int_{\partial V} \boldsymbol{w} \cdot \boldsymbol{n} \, dA = \int_{\partial V} x n_x \, dA, \tag{2.54}$$

hence

$$V = \int_{\partial V} x n_x \, dA. \tag{2.55}$$

By choosing $\boldsymbol{w} = [0, y, 0]^{\mathrm{T}}$ or $\boldsymbol{w} = [0, 0, z]^{\mathrm{T}}$, the remaining equations can likewise be proved.

EXAMPLE 2.8

A scalar field $\Phi(\boldsymbol{x}) = x_1^2 + 3x_2x_3$ describes some physical quantity (i.e., total potential energy). Show that the directional derivative of Φ in the direction $\boldsymbol{u} = \frac{1}{\sqrt{3}}(1, 1, 1)^T$ at the position $\boldsymbol{x} = (2, -1, 0)^T$ is $\frac{1}{\sqrt{3}}$.

Solution

Using Equation (2.10) gives

$$D\Phi(\boldsymbol{x})[\boldsymbol{u}]$$

$$= \left.\frac{d}{d\epsilon}\right|_{\epsilon=0} \left[(x_1 + \epsilon u_1)^2 + 3(x_2 + \epsilon u_2)(x_3 + \epsilon u_3)\right] \tag{2.56a}$$

$$= \left.\left(2(x_1 + \epsilon u_1)u_1 + 3(x_2 + \epsilon u_2)u_3 + 3(x_3 + \epsilon u_3)u_2\right)\right|_{\epsilon=0} \tag{2.56b}$$

$$= 2x_1u_1 + 3x_2u_3 + 3x_3u_2. \tag{2.56c}$$

Substituting the given values for $\boldsymbol{x}$ and $\boldsymbol{u}$ yields $D\Phi(\boldsymbol{x})[\boldsymbol{u}] = \frac{1}{\sqrt{3}}$.

EXAMPLE 2.9

Given the second-order tensor $\boldsymbol{A}$, obtain the directional derivative of the expression $\boldsymbol{A}^3 = \boldsymbol{AAA}$ in the direction of an arbitrary increment $\boldsymbol{U}$ of $\boldsymbol{A}$.

Solution

Applying the product rule for the directional derivative gives

$$D(\boldsymbol{A}^3)[\boldsymbol{U}] = D\boldsymbol{A}[\boldsymbol{U}]\boldsymbol{A}^2 + \boldsymbol{A}D\boldsymbol{A}[\boldsymbol{U}]\boldsymbol{A} + \boldsymbol{A}^2 D\boldsymbol{A}[\boldsymbol{U}], \tag{2.57}$$

where

$$D\boldsymbol{A}[\boldsymbol{U}] = \left.\frac{d}{d\epsilon}\right|_{\epsilon=0} (\boldsymbol{A} + \epsilon\boldsymbol{U}) \tag{2.58a}$$

$$= \boldsymbol{U}, \tag{2.58b}$$

giving

$$D(\boldsymbol{A}^3)[\boldsymbol{U}] = \boldsymbol{U}\boldsymbol{A}^2 + \boldsymbol{A}\boldsymbol{U}\boldsymbol{A} + \boldsymbol{A}^2\boldsymbol{U}. \tag{2.59}$$

EXAMPLE 2.10

Given $\boldsymbol{S}^{-1}\boldsymbol{S} = \boldsymbol{I}$ and $D\boldsymbol{I}[\boldsymbol{U}] = \boldsymbol{0}$, where $\boldsymbol{S}$ is a second-order tensor, show that $D(\boldsymbol{S}^{-1})[\boldsymbol{U}] = -\boldsymbol{S}^{-1}\boldsymbol{U}\boldsymbol{S}^{-1}$.

Solution

$$D(\boldsymbol{S}^{-1}\boldsymbol{S})[\boldsymbol{U}] = D\boldsymbol{S}^{-1}[\boldsymbol{U}]\boldsymbol{S} + \boldsymbol{S}^{-1}D\boldsymbol{S}[\boldsymbol{U}] \tag{2.60a}$$

$$= \boldsymbol{0}, \tag{2.60b}$$

$$D(\boldsymbol{S}^{-1})[\boldsymbol{U}] = -\boldsymbol{S}^{-1}\boldsymbol{U}\boldsymbol{S}^{-1}. \tag{2.60c}$$

EXAMPLE 2.11

If the second-order tensors $\boldsymbol{S} = \boldsymbol{S}^T$ and $\boldsymbol{W}^T = -\boldsymbol{W}$, show that $\boldsymbol{S} : \boldsymbol{W} = 0$ (i.e., $\mathrm{tr}(\boldsymbol{SW}) = 0$).

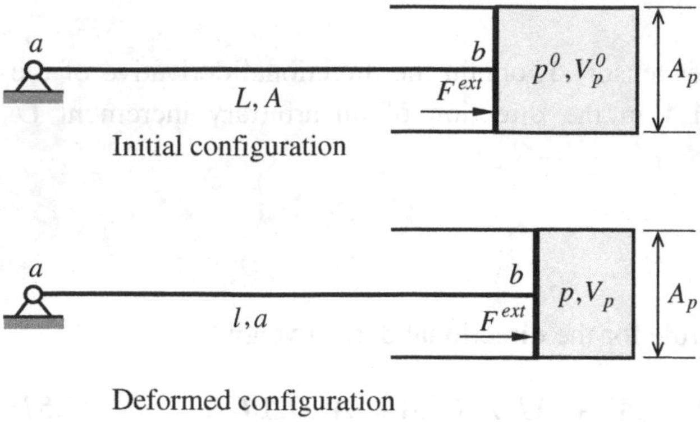

FIGURE 2.2 Truss – piston system

Solution

Recalling the relation between the double product and the trace given by Equation (2.7a,b)

$$\boldsymbol{S} : \boldsymbol{W} = \text{tr}(\boldsymbol{S}^T \boldsymbol{W}) \tag{2.61a}$$

$$= \text{tr}(\boldsymbol{S}\boldsymbol{W}) \tag{2.61b}$$

$$= \text{tr}(\boldsymbol{S}\boldsymbol{W}^T) \tag{2.61c}$$

$$= -\text{tr}(\boldsymbol{S}\boldsymbol{W}). \tag{2.61d}$$

For Equation (2.61b) to be equal to Equation (2.61d), $\text{tr}(\boldsymbol{S}\boldsymbol{W}) = 0$.

EXAMPLE 2.12

This provides an interesting example of the use of the directional derivative to find an equilibrium equation and the corresponding tangent stiffness term. Some material from textbook Chapter 3 is referred to in this example; however, the example is introduced here mainly to illustrate the use of the directional derivative. This simple model is analogous to the structural problem of a high altitude pumpkin balloon in which the axial rod represents the fabric and cabling enclosing a constant mass of gas. As the balloon rises, the external force provided by the atmospheric pressure decreases and the balloon expands. Figure 2.2 represents two configurations of a one-dimensional structural system comprised of a truss member joining nodes a and b of initial length L and cross-sectional area A connected to a piston chamber of constant section A_p filled with a gas. An external force F^{ext} is applied at node b. The constant mass of gas enclosed in the piston is considered to satisfy the Boyle's law which states that

$$p\,V_p = p^0\,V_p^0 = K; \quad K \text{ is a constant}$$

where p and V_p denote the pressure and the volume of the gas within the chamber in the deformed configuration, respectively. This example employs the strain energy per unit initial volume given in textbook Equation (3.17) as $\psi = E(\ln\lambda)^2/2$ where E is a Young's modulus type constitutive term and λ is the stretch of the axial rod. Consequently, the total energy functional for the overall system can be written as a function of the spatial position of point b denoted by the deformed length l and the initial volume AL as follows:

$$\begin{aligned}\Pi(l) &= \Pi^{\text{truss}}(l) + \Pi^{\text{piston}}(l) - \Pi^{\text{ext}}\\ &= \frac{1}{2}EAL\,(\ln\lambda)^2 - \int_{V_p} p(V)dV - F^{ext}l.\end{aligned} \tag{2.62}$$

in which $\lambda = \frac{l}{L}$ is the stretch ratio of the truss member.
(a) Obtain the stationary point of the above energy functional and derive the principal of virtual work.
(b) Obtain the equilibrium equation at node b and show that it is indeed a nonlinear equation.
(c) Obtain the tangent stiffness matrix required for a Newton–Raphson algorithm after suitable linearization.
(d) Explain whether the inclusion of the piston chamber increases or decreases the stiffness of the truss member.

Solution

(a) From Equation (2.62) the derivative of $\Pi^{\text{truss}}(l)$ in the direction δl (which could be real or virtual) to be found as[1]

$$\begin{aligned}D\Pi^{\text{truss}}(l)[\delta l] &= EAL\ln\left(\frac{l}{L}\right) D\left(\ln\left(\frac{l}{L}\right)\right)[\delta l]\\ &= EAL\ln\left(\frac{l}{L}\right)\frac{L}{l}D\left(\frac{l}{L}\right)[\delta l]\end{aligned}$$

[1] Based on the experience of previous examples given in this chapter, the explicit introduction of the derivative with respect to ϵ in the directional derivative is omitted.

$$= EAL \ln\left(\frac{l}{L}\right)\left(\frac{L}{l}\right)\frac{1}{L}\delta l$$

$$= \frac{EAL}{l} \ln\left(\frac{l}{L}\right) \delta l. \tag{2.63}$$

Before finding the directional derivative of the piston component of the total energy, it is necessary to find the deformed volume of the gas as a function of the deformed length l as

$$V_p(l) = V_p^0 - A_p(l - L). \tag{2.64}$$

From Equation (2.62) the gas pressure as a function of l is

$$p(l) = \frac{p^0 V_p^0}{V_p^0 - A_p(l - L)}. \tag{2.65}$$

The directional derivative of the piston component of the total energy with respect to δl can now be found as

$$\begin{aligned} D\Pi^{\text{piston}}(l)[\delta l] &= \frac{\partial \Pi^{\text{piston}}}{\partial V_p} DV_p(l)[\delta l] \\ &= -p(l) DV_p(l)[\delta l]. \end{aligned} \tag{2.66}$$

where from Equation (2.64)

$$DV_p(l)[\delta l] = -A_p \delta l. \tag{2.67}$$

Finally, the derivative of the external total energy component $F^{\text{ext}} l$ is simply

$$D\Pi^{\text{ext}}(l)[\delta l] = F^{\text{ext}} \delta l. \tag{2.68}$$

Assembling the complete directional derivative from Equations (2.63, 2.65, 2.66, 2.68) and noting the negative terms in Equation (2.62) yields the stationary condition as

$$D\Pi(l)[\delta l] = \left(EA\frac{L}{l} \ln\frac{l}{L} + \frac{A_p p^0 V_p^0}{V_p^0 - A_p(l - L)} - F^{\text{ext}} \right) \delta l = 0. \tag{2.69}$$

The three terms inside the parentheses in the above equation are all forces acting at point b. Consequently, treating δl now as a virtual change in length of the rod, Equation (2.69) can be identified as the virtual work expression of equilibrium as

$$\delta W(l, \delta l) = \left(T^{\text{truss}}(l) + T^{\text{piston}} - F^{\text{ext}}\right) \delta l = 0. \tag{2.70}$$

(b) In Equation (2.70) δl although virtual is not zero, hence the equilibrium equation can be established as

$$R(l) = T^{\text{truss}}(l) + T^{\text{piston}}(l) - F^{\text{ext}} = 0. \tag{2.71}$$

(c) The tangent stiffness term used in a Newton–Raphson procedure is found by linearizing the equilibrium equation. This is achieved by finding the directional derivative of Equation (2.71) in the direction of a real change u in l to give

$$DR(l)[u] = DT^{\text{truss}}(l)[u] + DT^{\text{piston}}[u] - DF^{\text{ext}}[u] = 0, \tag{2.72}$$

where

$$\begin{aligned} DT^{\text{truss}}(l)[u] &= -\frac{EAL}{l^2} \ln\frac{l}{L} D(l)[u] + \frac{EAL}{l}\left(\frac{L}{l}\right) D\left(\frac{l}{L}\right)[u] \\ &= \left(-\frac{EAL}{l^2} \ln\frac{l}{L} + \frac{EAL}{l^2}\right) u \\ &= \frac{EAL}{l^2}\left(1 - \ln\frac{l}{L}\right) u \end{aligned} \tag{2.73}$$

$$\begin{aligned} DT^{\text{piston}}(l)[u] &= \frac{-A_p p^0 V_p^0}{(V^0 - A_p(l-L))^2} D\left(V^0 - A_p(l-L)\right)[u] \\ &= \left(\frac{A_p^2 p^0 V_p^0}{(V^0 - A_p(l-L))^2}\right) u. \end{aligned} \tag{2.74}$$

Consequently from Equations (2.73, 2.74) the tangent stiffness is

$$K = \frac{EAL}{l^2}\left(1 - \ln\frac{l}{L}\right) + \frac{A_p^2 p^0 V_p^0}{(V^0 - A_p(l-L))^2}. \tag{2.75}$$

(d) From the final term in the above equation it can be seen that the stiffness of the truss is increased by the presence of the piston chamber.

CHAPTER THREE

ANALYSIS OF THREE-DIMENSIONAL TRUSS STRUCTURES

The two- and three-dimensional truss examples presented in this chapter demonstrate the complex and often unexpected load deflection behavior exhibited when, in particular, geometrical nonlinearity is included in structural analysis. Each point on the various graphs shown below represents an equilibrium configuration; however, these configurations may be structurally stable or unstable. For a chosen load it can be observed that the structure can be in a variety of equilibrium configurations. For most structures subjected to "in service" loadings, this is clearly unacceptable (not to say alarming), nevertheless such analysis can indicate possible collapse scenarios. While the points on a load deflection graph refer to equilibrium configurations, it must not be assumed that connecting adjacent points necessarily represents smooth continuity of the motion of the structure as the loading changes. However, such smooth motion is likely to be the case if a large number of load increments are employed in the solution, but it cannot be guaranteed.

A situation where a small change in load leads to a dramatic change in configuration is known as "snap-through" behavior. There are "structures" that rely on snap-through behavior to fulfill a useful function. Indeed such structures are vastly more numerous than everyday structures; for example, a shampoo container cap when opened carefully will suddenly "flick" into a fully opened position. A child's hair clip often employs snap-through behavior to lock into position, while perhaps the most common item is the simple light switch. Unfortunately, shallow dome structures can exhibit snap-through characteristics with disastrous consequences.

Unlike linear analysis, controlling nonlinear finite element solutions requires experience. New users to the FLagSHyP program (available free at www.flagshyp.com) can often be frustrated when the solution fails, produces unexpected results, or produces results distant from the region of interest. Satisfactory results can only be obtained by careful adjustment of the control parameters which comprise the final line of data. These are given below and are fully discussed in the second edition of the textbook, *Nonlinear Continuum Mechanics for Finite Element Analysis.*

Users are advised to change the control data to see the effect and gain experience (as have many students of this subject!).

`nincr,xlmax,dlamb,`
`miter,cnorm,searc,`
`arcln,incout,itarget`
`nwant, iwant`
	1	Solution control parameters:
		`nincr:` number of load/ displacement increments
		`xlmax:` maximum value of load-scaling parameter
		`dlamb:` load parameter increment
		`miter:` maximum allowed number of iterations per increment
		`cnorm:` convergence tolerance
		`searc:` line-search parameter (if 0.0 not in use)
		`arcln:` arc-length parameter (if 0.0 not in use)
		`incout:` output counter (e.g., for every 5th increment, `incout=5`)
		`itarget:` target iterations per increment (see note below)*
		`nwant:` single output node (0 if not used)
		`iwant:` output degree of freedom at `nwant` (0 if not used) (see note 5)*

* See user instructions in Section 10.2 of the main text.

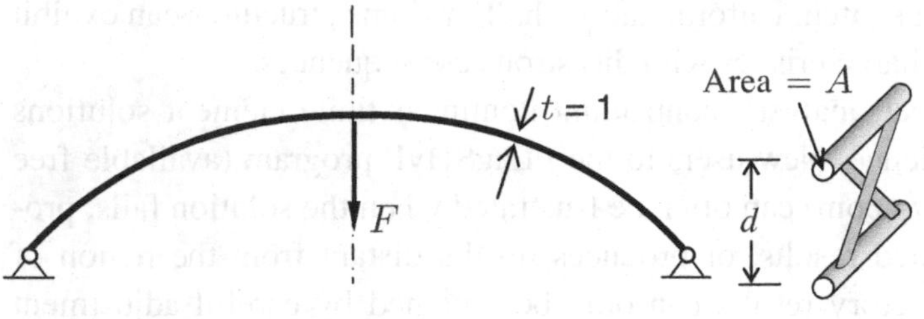

FIGURE 3.1 Arch

EXAMPLE 3.1: Textbook Exercise 3.1

Run the simple single degree of freedom example given in Section 3.6.1, Figure 3.8. A high value of the yield stress will ensure that the truss remains elastic.

Solution

```
one truss element
truss2
2
1 7 0.0 0.0 0.0
2 5 100.0 100.0 0.0
1
1 1 1 2
1
1 2
0.0 210000.0 0.3 1.0 1.0e+10 1.0
1 0 0 0.0 0.0 0.0
2 0.0 -1.0 0.0
500 40000.0 0.01 10 1.0E-6 0.0 -0.50 5 5 2 2
```

EXAMPLE 3.2: Textbook Exercise 3.2

Analyze the arch shown in text Figure 3.10 (Figure 3.1 above). The radius is 100, the height is 40 and the half span is 80. The cross-sectional area is

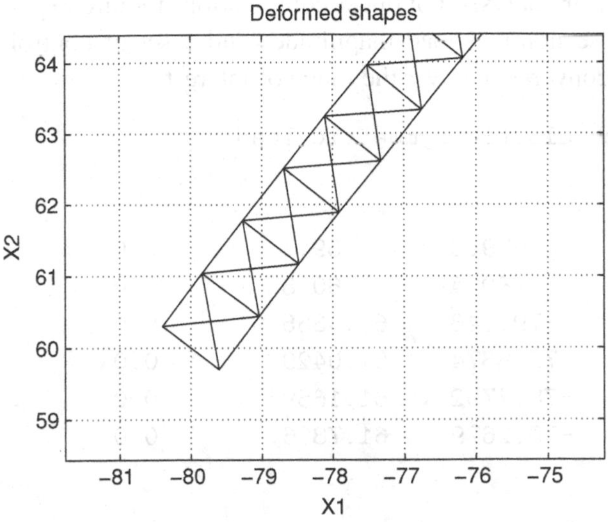

FIGURE 3.2 Arch: left support

1×1. Young's modulus is 10^7 and Poisson's ratio is $\nu = 0.3$. The figure shows how the arch can be represented as a truss where, by ignoring the cross members, the second moment of area of the arch, $I = 1/12$, can be approximated by the top and bottom truss members, where $I = 2(A(t/2)^2)$. Plot the central load vertical deflection curve. Slight imperfections in the symmetry of the geometry may caused unsymmetric deformations, otherwise these can be initiated by a very small horizontal load being placed with the vertical load.

Solution

The area of the top and bottom chords of the truss is calculated as $a = 1/6$ and the area of the interconnecting members is chosen as $a = 1/12$. Top and bottom chords have material number 1 and all others material number 2. Figure 3.2 shows the truss configuration in the x–y plane with the pin support (boundary code 7) being on the lower end node. Intermediate nodes are all restrained in the out of plane z direction using boundary code 4. Only partial coordinate and element data is shown below as there are 402 nodes and 1001 elements. A variable arc length is employed.

Note that, unlike linear analysis, nonlinear computations require experience in order to choose nominal load magnitudes and a set of control parameters that ensure convergence over the range of interest.

```
Circular truss arch elastic symmetric load
truss2
402
         1    7      -79.6        59.7          0.0
         2    4      -80.4        60.3          0.0
         3    4    -79.043     60.4355          0.0
         4    4   -79.8374     61.0429          0.0
         5    4   -78.4792     61.1659          0.0
         6    4   -79.2679     61.7806          0.0
         .
         .
         .
       397    4    78.4792     61.1659          0.0
       398    4    79.2679     61.7806          0.0
       399    4     79.043     60.4355          0.0
       400    4    79.8374     61.0429          0.0
       401    7       79.6        59.7          0.0
       402    4       80.4        60.3          0.0
1001
         1         2         1         2
         2         1         2         4
         3         1         1         3
         4         2         1         4
         5         2         2         3
         6         2         3         4
         7         1         4         6
         8         1         3         5
         9         2         3         6
        10         2         4         5
         .
         .
         .
       990         2       396       397
       991         2       397       398
```

```
        992         1         398        400
        993         1         397        399
        994         2         397        400
        995         2         398        399
        996         2         399        400
        997         1         400        402
        998         1         399        401
        999         2         399        402
       1000         2         400        401
       1001         2         401        402
2
1 2
0.0 10000000.0 0.3 0.166666667 250000000.0 1.0
2 2
0.0 10000000.0 0.3 0.083333333 250000000.0 1.0
1 0 0 0.0 0.0 0.0
202 0.0 -10.0 0.0
250 25000.0 0.1 100 1.0E-6 0.0 1.0 1 5 202 2
```

Figure 3.3 shows the load deflection (not position) behavior at the central top node, where negative loads (i.e., downward) are shown positive and points are joined with straight lines. Observe that all points on the curves in Figure 3.3 represent positions of equilibrium, some stable, many unstable. When running the example the appearance of solver warnings in an otherwise successful computation usually indicates regions of unstable equilibrium.

The convoluted equilibrium paths are similar to those achieved in the paper by H. B. Harrison entitled *Post-buckling behavior of elastic circular arches* in the Proc. Instn. Civ. Engrs, Part 2, 1978, 65, June, 283–98. Figure 3.4 shows equilibrium configurations at increments 23 and 40 corresponding to loads of 1259.6 (downward, negative on output) and 2930.1 (upward, positive on output).

Figure 3.3 clearly shows that at a given load there is the possibility of a number of equilibrium configurations. This is demonstrated in Figure 3.5 where some unexpected configurations occur at about a downward load of 5000. To initiate an unsymmetric deformation, the load and controls data lines are changed to

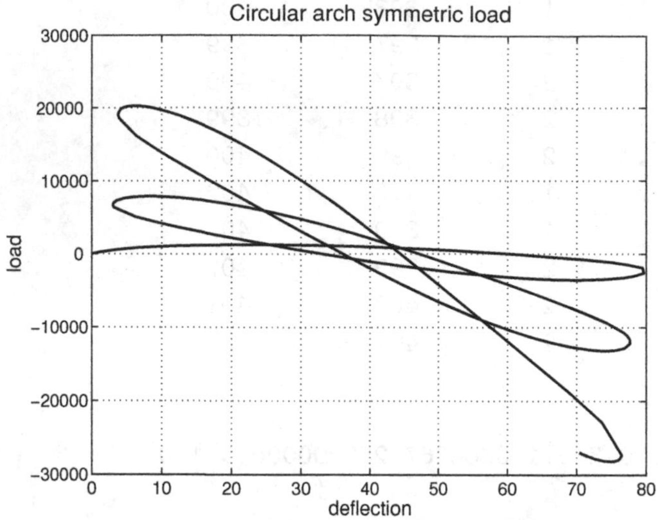

FIGURE 3.3 Arch: symmetric load deflection behavior

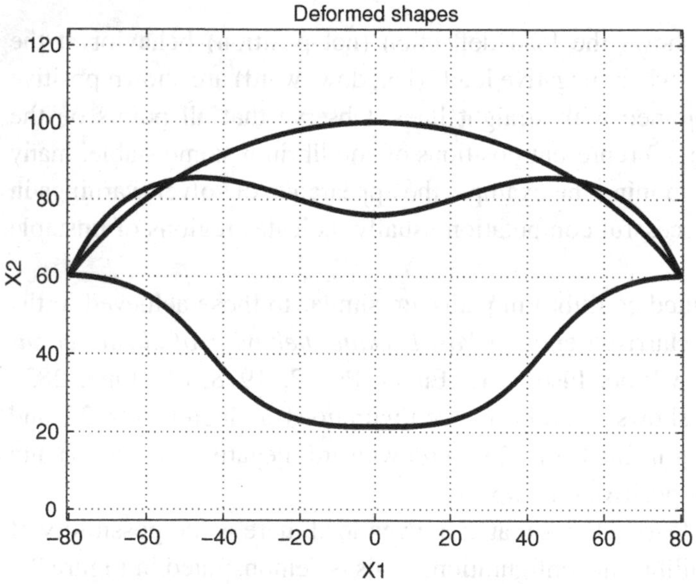

FIGURE 3.4 Arch: symmetric equilibrium configurations 1

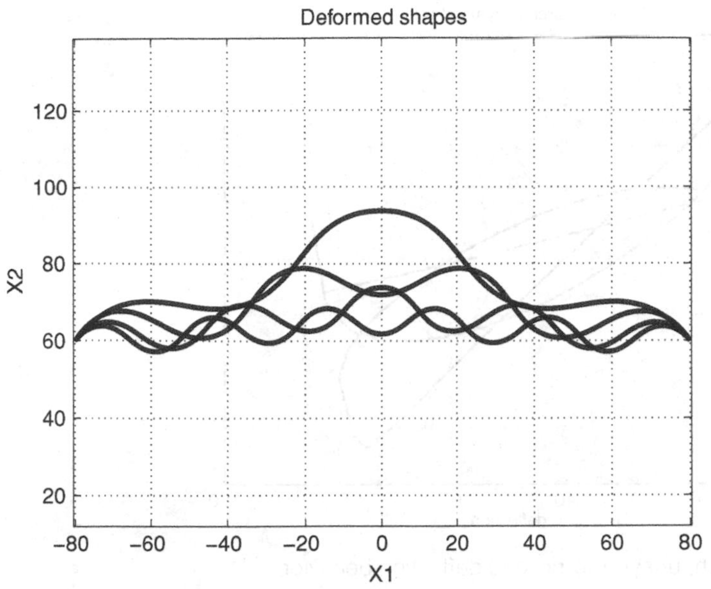

FIGURE 3.5 Arch: symmetric equilibrium configurations 2

```
202 0.01 -10.0 0.0
250 25000.0 0.1 100 1.0E-6 0.0 1.0 1 5 202 2
```

which results in the emergence of a bifurcation in the equilibrium path at a load of about 1000, as shown in Figure 3.6. Note that the solution failed to converge after increment 212, which can probably be overcome using a fixed arc length.

Various unsymmetric configurations are shown in Figure 3.7.

EXAMPLE 3.3: Textbook Exercise 3.3

Analyze the shallow trussed dome shown in textbook Figure 3.11. The outer radius is 50 and height 0, the inner radius is 25 and height 6.216, and the apex height is 8.216. The cross-sectional area of each truss member is unity. Textbook Figure 3.11 is approximate in that the apparent major triangles spanning the outer circle do not have straight sides as shown. Young's modulus is 8×10^7 and Poisson's ratio is 0.5, indicating incompressible behavior. Figure 3.8 shows the initial shape.

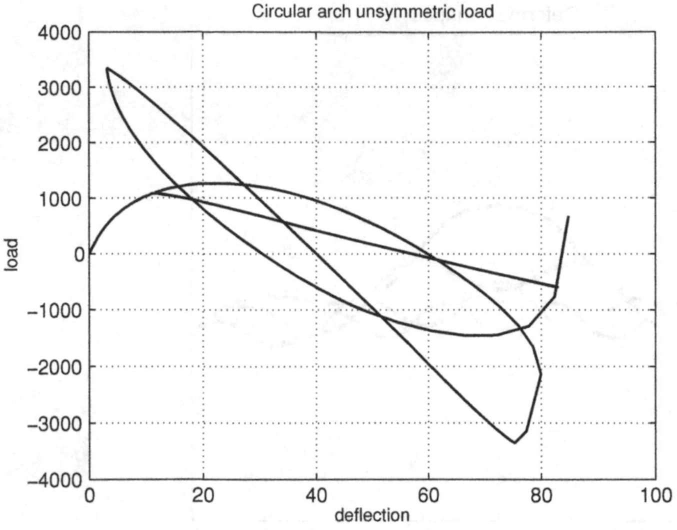

FIGURE 3.6 Arch: unsymmetric load deflection behavior

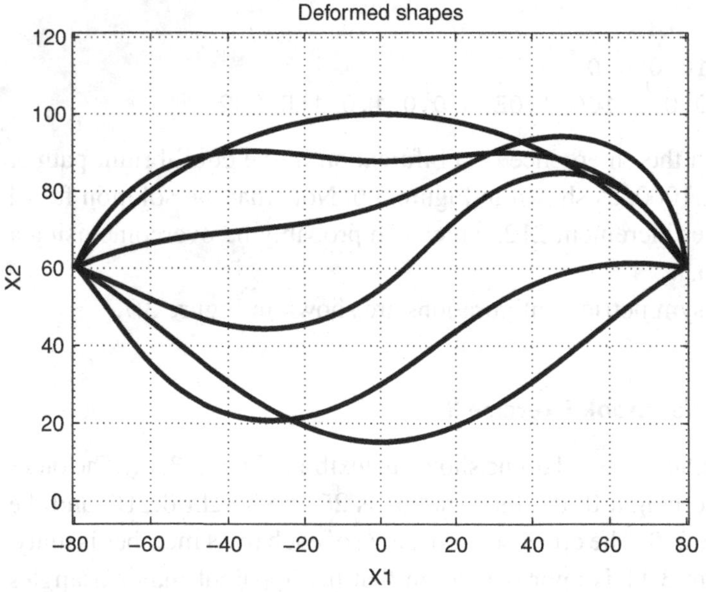

FIGURE 3.7 Arch: unsymmetric equilibrium configurations

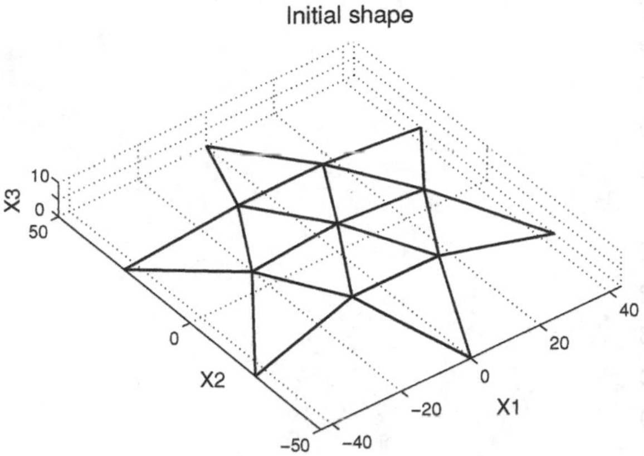

FIGURE 3.8 Shallow dome: initial shape

Solution

```
3D shallow dome example
truss2
  13
     1  7     0.0      50.0     0.0
     2  7  -43.3013  25.0     0.0
     3  7  -43.3013 -25.0     0.0
     4  7     0.0    -50.0     0.0
     5  7   43.3013 -25.0     0.0
     6  7   43.3013  25.0      0.0
     7  0   -12.5     21.6506  6.216
     8  0   -25.0      0.0     6.216
     9  0   -12.5    -21.6506  6.216
    10  0    12.5    -21.6506  6.216
    11  0    25.00     0.0     6.216
    12  0    12.5     21.6506  6.216
    13  0     0.0      0.0     8.216
  24
     1    1    1  12
     2    1    1   7
     3    1    2   7
     4    1    2   8
     5    1    3   8
```

```
    6    1    3   9
    7    1    4   9
    8    1    4  10
    9    1    5  10
   10    1    5  11
   11    1    6  11
   12    1    6  12
   13    1    7  12
   14    1    7   8
   15    1    8   9
   16    1    9  10
   17    1   10  11
   18    1   11  12
   19    1   12  13
   20    1    7  13
   21    1    8  13
   22    1    9  13
   23    1   10  13
   24    1   11  13
1
1  2
0.0     80.0E+06    0.5     1.0 10.0E+06 1.0
1  0    0      0.0      0.0       0.0
13     0.0       0.0  -100.0
200  1.0E+06  1.0  20  1.0E-06  0.0  -1.0 5  5  13  3
```

Plot the vertical downward load deflection behavior at the apex. A fixed arc length was used to achieve the equilibrium points shown on Figure 3.9 (not all 200 increments shown). This is a good example of snap-through behavior. The equilibrium path is very convoluted, but upon examination the corresponding dome shapes are perfectly reasonable.

EXAMPLE 3.4: Textbook Exercise 3.4

Run the trussed frame example given in textbook Figure 3.9, initially as shown and then with clamped supports. The cross-sectional area is 6, giving a truss member area of 1. Typical equilibrium configurations are given in Figure 3.10 and Figure 3.11 shows the load deflection (equilibrium path).

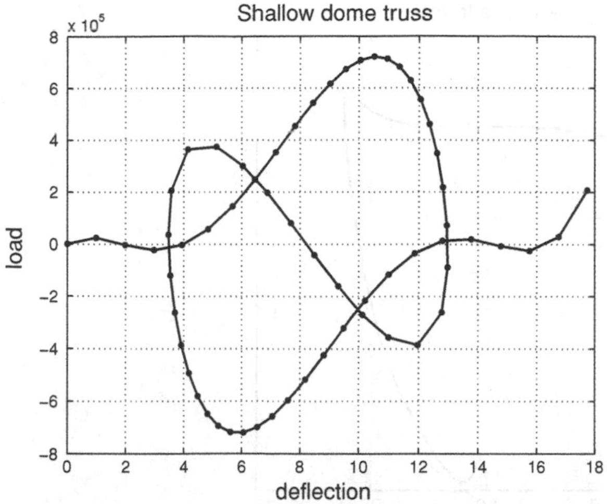

FIGURE 3.9 Shallow dome: load deflection behavior at the apex

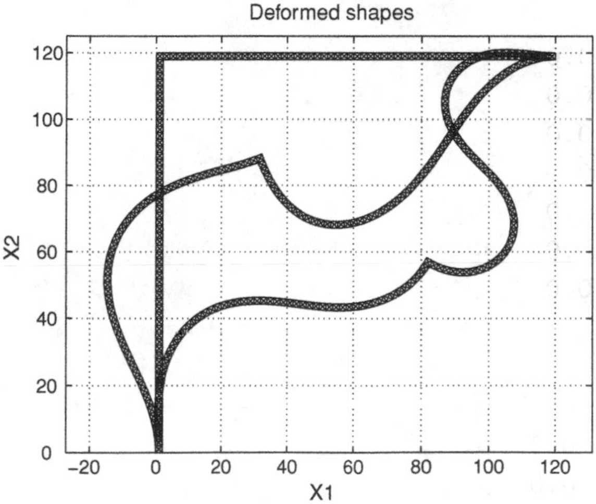

FIGURE 3.10 Clamped Lee's frame: equilibrium configurations

Solution

```
Clamped Lee's frame with truss elements elastic case
truss2
 240
 1   7 0.0    0.0   0.0
```

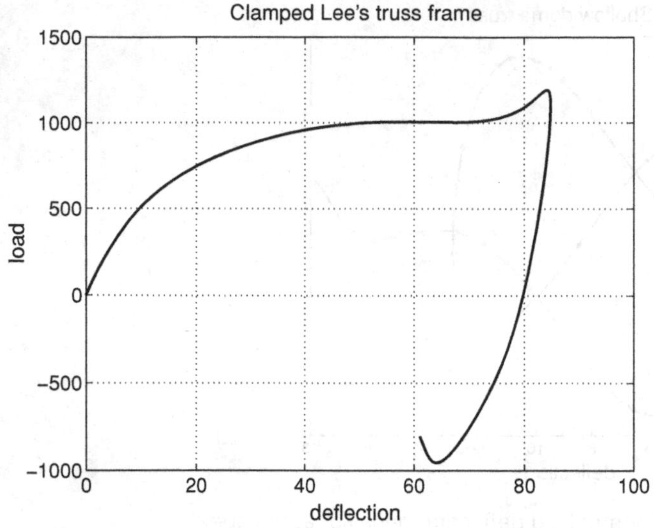

FIGURE 3.11 Clamped Lee's frame: load deflection behavior

```
2    7 2.0    0.0   0.0
3    4 0.0    2.0   0.0
4    4 2.0    2.0   0.0
5    4 0.0    4.0   0.0
6    4 2.0    4.0   0.0
7    4 0.0    6.0   0.0
8    4 2.0    6.0   0.0
.
.
.
233 4 114.0 118.0   0.0
234 4 114.0 120.0   0.0


235 4 116.0 118.0   0.0
236 4 116.0 120.0   0.0
237 4 118.0 118.0   0.0
238 4 118.0 120.0   0.0
239 7 120.0 118.0   0.0
240 7 120.0 120.0   0.0
```

```
596
1   1     1          2
2   1     1          3
3   1     2          4
4   1     1          4
5   1     2          3
.
.
.
590 1   236        237
591 1   237        238
592 1   237        239
593 1   238        240
594 1   237        240
595 1   238        239
596 1   239        240
1
1 2
0.0 210000.0 0.3 1.0 10.0E+06 1.0
1 0 0 0.0 0.0 0.0
144 0.0 -100.0 0.0
500 25000.0 0.1 100 1.0E-6 0.0 -10.0 5 5 144 2
```

EXAMPLE 3.5

This example, devised by Crisfield,[1] demonstrates snap-back behavior. The layout is shown in Figure 3.12 where elements 1, 2, and 4 support an effectively rigid element 3. In the original example, elements 1, 2, and 4 were linear springs; consequently the lengths and material properties of these elements are chosen to model linear springs in which the initial stress term is negligible.

[1] Crisfield, M. A., *Non-linear Finite Element Analysis of Solids and Structures*, Volume 1, Wiley, 1991, p. 100.

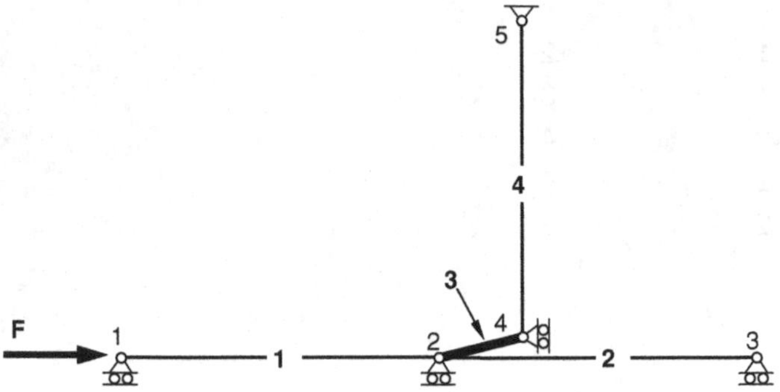

FIGURE 3.12 Crisfield's snap-back problem – configuration

The length of elements 1, 2, and 4 is 10^7 and sloping element 3 is approximately 2500. Other material properties are such that

$$\left(\frac{EA}{L}\right)_1 = 1.00; \quad \left(\frac{EA}{L}\right)_2 = 0.25; \quad \left(\frac{EA}{L}\right)_4 = 1.50 \qquad (3.1)$$

A high value, $1.0e10$, is chosen for the yield stress to ensure elasticity.

Solution

```
Crisfield's snap-back problem
truss2
 5
 1 6 -10000000.0 0.0    0.0
 2 6           0.0 0.0    0.0
 3 7 10000000.0  0.0    0.0
 4 3    2500.0      0.0  25.0
 5 7 2500.0 0.0 10000025.0
 4
 1 1 1 2
 2 2 2 3
 3 3 2 4
 4 4 4 5
 4
 1 2
 0.0 10000000.0 0.0 1.00 1.0e+10 1.0
```

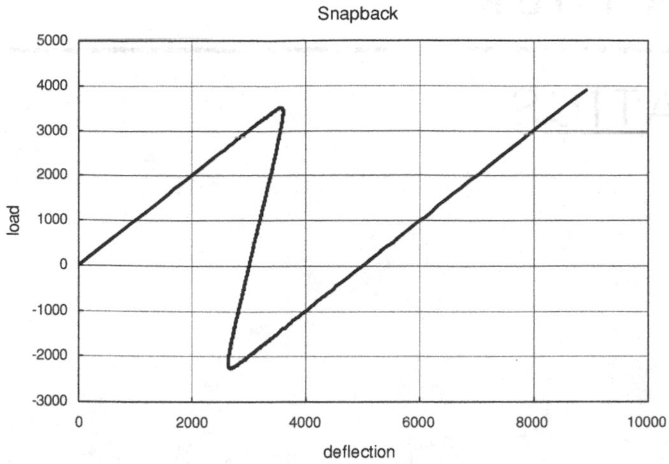

FIGURE 3.13 Snap-back problem – horizontal load deflection for node 1

```
2 2
0.0 10000000.0 0.0 0.25 1.0e+10 1.0
3 2
0.0 50000000.0 0.0 1.0 1.0e+10 1.0
4 2
0.0 10000000.0 0.0 1.5 1.0e+10 1.0
1 0 0 0.0 0.0 0.0
1 100.0 0.0 0.0
1000 100.0 1.0 20 1.0e-06 0.0 1.0 1 3 1 1
```

Figure 3.13 shows the snap-back behavior where the central portion of the equilibrium path shows both a reduction in load and displacement.

CHAPTER FOUR

KINEMATICS

This chapter provides examples involving various aspects of finite deformation kinematics. The examples are presented in a pedagogical order and include additional examples not in the textbook.

Equation summary

Notation for transpose of the inverse of a square matrix $\boldsymbol{A}$

$$\boldsymbol{A}^{-\mathrm{T}} = (\boldsymbol{A}^{-1})^{\mathrm{T}} = (\boldsymbol{A}^{\mathrm{T}})^{-1}. \tag{4.1}$$

Third invariant of a second-order tensor [2.54]

$$III_{\boldsymbol{S}} = \det \boldsymbol{S} = \det[\boldsymbol{S}]. \tag{4.2}$$

Push forward of elemental material vector [4.10,12]

$$d\boldsymbol{x} = \boldsymbol{F} d\boldsymbol{X}. \tag{4.3}$$

Green's strain tensor [4.18b]

$$\boldsymbol{E} = \frac{1}{2}(\boldsymbol{C} - \boldsymbol{I}) \; ; \quad \boldsymbol{C} = \boldsymbol{F}^{\mathrm{T}}\boldsymbol{F}. \tag{4.4}$$

Rotation tensor [4.27]

$$\boldsymbol{R} = \boldsymbol{F}\boldsymbol{U}^{-1}. \tag{4.5}$$

Right Cauchy–Green tensor [4.30]

$$C = \sum_{\alpha=1}^{3} \lambda_\alpha^2 \, N_\alpha \otimes N_\alpha. \tag{4.6}$$

Material stretch tensor [4.31]

$$U = \sum_{\alpha=1}^{3} \lambda_\alpha \, N_\alpha \otimes N_\alpha. \tag{4.7}$$

Push forward of elemental material vector in terms of spatial stretch tensor and rotation tensor [4.34]

$$d\boldsymbol{x} = \boldsymbol{F} d\boldsymbol{X} = \boldsymbol{V}(\boldsymbol{R} d\boldsymbol{X}). \tag{4.8}$$

Stretch [4.42]

$$\lambda_1 = \frac{dl_1}{dL_1}. \tag{4.9}$$

Alternative relation between spatial and material vector triads [4.44a]

$$\boldsymbol{F}\boldsymbol{N}_\alpha = \lambda_\alpha \boldsymbol{n}_\alpha. \tag{4.10}$$

Relation between spatial and matcrial vector triads [4.44b]

$$\boldsymbol{F}^{-\mathrm{T}}\boldsymbol{N}_\alpha = \frac{1}{\lambda_\alpha}\boldsymbol{n}_\alpha. \tag{4.11}$$

Elemental volume ratio [4.57]

$$dv = J dV \,;\; J = \det \boldsymbol{F}. \tag{4.12}$$

Distortional component of the right Cauchy–Green tensor [4.65]

$$\hat{\boldsymbol{C}} = (\det \boldsymbol{C})^{-1/3}\boldsymbol{C}. \tag{4.13}$$

Push forward of elemental material area vector [4.68]

$$d\boldsymbol{a} = J\boldsymbol{F}^{-\mathrm{T}} d\boldsymbol{A}. \tag{4.14}$$

Linearized deformation gradient [4.69]

$$D\boldsymbol{F}(\boldsymbol{\phi}_t)[\boldsymbol{u}] = (\boldsymbol{\nabla}\boldsymbol{u})\boldsymbol{F}. \tag{4.15}$$

Linearized Green's strain [4.72]

$$D\boldsymbol{E}[\boldsymbol{u}] = \boldsymbol{\phi}_*^{-1}[\boldsymbol{\varepsilon}] = \boldsymbol{F}^{\mathrm{T}}\boldsymbol{\varepsilon}\boldsymbol{F}. \tag{4.16}$$

Linearized right Cauchy–Green tensor [4.74a]

$$DC[u] = 2F^T \varepsilon F. \tag{4.17}$$

Linearized volume change [4.77]

$$DJ[u] = J\,\mathrm{tr}\varepsilon. \tag{4.18}$$

Velocity of a particle [4.80]

$$v(X, t) = \frac{\partial \phi(X, t)}{\partial t}. \tag{4.19}$$

Velocity gradient tensor [4.93]

$$l = \dot{F} F^{-1}. \tag{4.20}$$

Material strain rate tensor [4.97]

$$\dot{E} = \tfrac{1}{2}\dot{C} = \tfrac{1}{2}(\dot{F}^{\mathrm{T}} F + F^{\mathrm{T}} \dot{F}). \tag{4.21}$$

Rate of deformation tensor [4.100a]

$$d = F^{-T} \dot{E} F^{-1}. \tag{4.22}$$

Rate of deformation tensor [4.101]

$$d = \tfrac{1}{2}(l + l^T). \tag{4.23}$$

EXAMPLE 4.1: Textbook Exercise 4.1

(a) For the uniaxial strain case find the Engineering, Green's, and Almansi strain in terms of the stretch λ_1.
(b) Using these expressions, show that when the Engineering strain is small, all three strain measures converge to the same value.

Solution

The uniaxial stretch is simply defined by Equation (4.9) as

$$\lambda_1 = dl_1/dL_1 \tag{4.24}$$

from which the various uniaxial strain measures given in Chapter 1 are:

Engineering strain:

$$\varepsilon_E = \frac{dl - dL}{dL} = \lambda_1 - 1; \tag{4.25}$$

Lagrangian or Green strain:

$$\varepsilon_G = \frac{1}{2}\left(\frac{dl^2 - dL^2}{dL^2}\right) = \frac{1}{2}(\lambda_1^2 - 1); \tag{4.26}$$

Almansi strain:

$$\varepsilon_A = \frac{1}{2}\left(\frac{dl^2 - dL^2}{dl^2}\right) = \frac{1}{2}\left(1 - \frac{1}{\lambda_1^2}\right). \tag{4.27}$$

From Equation (4.25), the stretch λ_1 can be expressed in terms of the Engineering strain as

$$\lambda_1 = 1 + \varepsilon_E, \tag{4.28}$$

enabling Green and Almansi strains to be rewritten as

$$\varepsilon_G = \frac{1}{2}(\lambda_1^2 - 1) = \varepsilon_E + \frac{1}{2}\varepsilon_E^2, \tag{4.29a}$$

$$\varepsilon_A = \frac{1}{2}\left(1 - \frac{1}{\lambda_1^2}\right) = \frac{1}{2}\left(\frac{\varepsilon_E^2 + 2\varepsilon_E}{\varepsilon_E^2 + 2\varepsilon_E + 1}\right). \tag{4.29b}$$

For small Engineering strain, higher-order terms can be neglected, yielding

$$\varepsilon_G = \varepsilon_E + \frac{1}{2}\varepsilon_E^2 \approx \varepsilon_E, \tag{4.30a}$$

$$\varepsilon_A = \frac{1}{2}\left(\frac{\varepsilon_E^2 + 2\varepsilon_E}{\varepsilon_E^2 + 2\varepsilon_E + 1}\right) \approx \frac{\varepsilon_E}{1} = \varepsilon_E. \tag{4.30b}$$

EXAMPLE 4.2

A continuum body, see Figure 4.1, undergoes a rigid body rotation θ about the origin defined by

$$\boldsymbol{x} = \boldsymbol{R}\boldsymbol{X}\ ; \quad \boldsymbol{R} = \begin{bmatrix} \cos\theta & -\sin\theta \\ \sin\theta & \cos\theta \end{bmatrix} \tag{4.31}$$

where $\boldsymbol{R}$ is a rotation matrix. In other words, material point $P(X_1, X_2)$ rotates to spatial point $p(x_1, x_2)$.

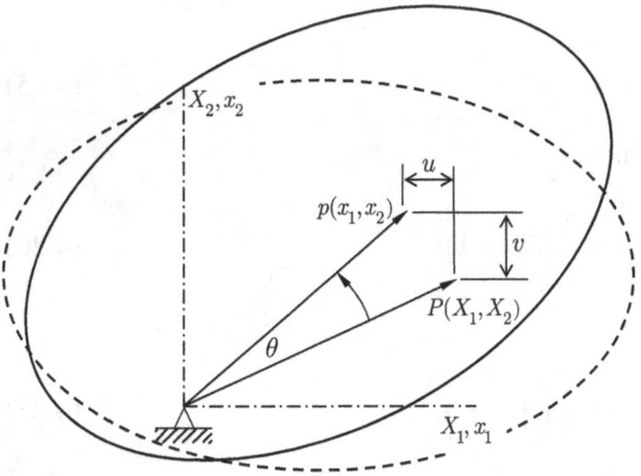

FIGURE 4.1 Rigid body rotation

(a) Demonstrate why the engineering or small strain tensor ε is not a valid measure of strain when the rotation θ is large.
(b) Demonstrate that Green's strain $\boldsymbol{E}$ is a valid measure of strain for the above motion regardless of the magnitude of θ.

Solution

(a) From Equation (4.31) the spatial coordinates of p are

$$x_1 = (\cos\theta)X_1 - (\sin\theta)X_2, \tag{4.32a}$$

$$x_2 = (\sin\theta)X_1 + (\cos\theta)X_2. \tag{4.32b}$$

The displacements $u(\boldsymbol{X})$ and $v(\boldsymbol{X})$ are

$$u(X_1, X_2) = -(X_1 - x_1) = (\cos\theta - 1)X_1 - (\sin\theta)X_2, \tag{4.33a}$$

$$v(X_1, X_2) = x_2 - X_2 = (\sin\theta)X_1 + (\cos\theta - 1)X_2. \tag{4.33b}$$

The relevant components of the engineering strain tensor ε are

$$\varepsilon_{xx} = \frac{\partial u_x}{\partial x}; \tag{4.34a}$$

$$\varepsilon_{yy} = \frac{\partial u_y}{\partial y}; \tag{4.34b}$$

$$\varepsilon_{xy} = \frac{1}{2}\left(\frac{\partial u_x}{\partial y} + \frac{\partial u_y}{\partial x}\right). \tag{4.34c}$$

From Equation (4.33) it is easily shown that the Engineering strain is not zero for the rigid body rotation; thus from Equation (4.31)

$$\boldsymbol{x} = \boldsymbol{R}\boldsymbol{X} = \boldsymbol{X} + \boldsymbol{u}, \tag{4.35}$$

giving

$$\begin{aligned}\boldsymbol{u} &= \boldsymbol{x} - \boldsymbol{R}^{\mathrm{T}}\boldsymbol{x}\\ &= (\boldsymbol{I} - \boldsymbol{R}^{\mathrm{T}})\boldsymbol{x}\\ &= \begin{bmatrix} 1-\cos\theta & -\sin\theta \\ \sin\theta & 1-\cos\theta \end{bmatrix}\boldsymbol{x},\end{aligned} \tag{4.36}$$

hence

$$\varepsilon_{xx} = \varepsilon_{yy} = (1-\cos\theta)\ ; \quad \varepsilon_{xy} = 0. \tag{4.37}$$

(b) For the rigid body case being considered, the deformation gradient tensor is identical to the rotation tensor, i.e., $\boldsymbol{F} = \boldsymbol{R}$ consequently from Equation (4.4) and noting that $\boldsymbol{R}^{\mathrm{T}} = \boldsymbol{R}^{-1}$

$$\boldsymbol{E} = \frac{1}{2}(\boldsymbol{R}^{\mathrm{T}}\boldsymbol{R} - \boldsymbol{I}) = \boldsymbol{0}. \tag{4.38}$$

Hence all components of Green's strain conform to zero strain as required of rigid body motion, as indeed does the Eulerian or Almansi strain tensor given by

$$\boldsymbol{e} = \frac{1}{2}(\boldsymbol{I} - \boldsymbol{b}^{-1})\ ; \quad \boldsymbol{b} = \boldsymbol{F}\boldsymbol{F}^{\mathrm{T}}. \tag{4.39}$$

EXAMPLE 4.3

A single 4-node isoparametric element can be used to illustrate the material and spatial coordinates used in finite deformation analysis.[1] The nondimensional coordinates ξ and η can be replaced by the material coordinates X_1 and X_2, giving the resulting single square element with initial dimensions 2×2 and centered at $X_1 = X_2 = 0$, see Figure 4.2. The shape functions are employed to define the spatial coordinates $\boldsymbol{x} = (x_1, x_2)^{\mathrm{T}}$ in terms of the material coordinates $\boldsymbol{X} = (X_1, X_2)^{\mathrm{T}}$ and $\boldsymbol{x}_a = (x_1, x_2)_a^{\mathrm{T}}$ where $\boldsymbol{x}_a$

[1] This example can be illustrated and calculated using the MATLAB program "polar_decomposition.m" which can be downloaded from the website www.flagshyp.com. This program can be used to explore the deformation using user-chosen spatial coordinates.

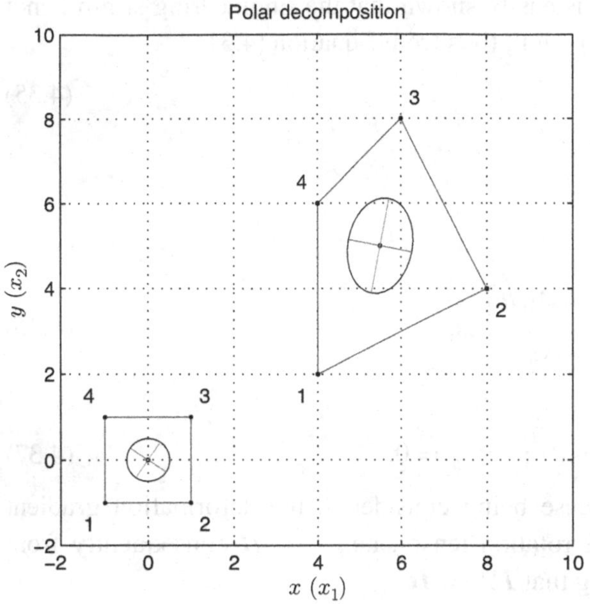

FIGURE 4.2 Polar decomposition

contains the spatial coordinates of nodes $a = 1, 4$ in the deformed configuration. Consequently the mapping $\boldsymbol{x} = \phi(\boldsymbol{X})$ is

$$\boldsymbol{x} = \phi(\boldsymbol{X}) = N_1\boldsymbol{x}_1 + N_2\boldsymbol{x}_2 + N_3\boldsymbol{x}_3 + N_4\boldsymbol{x}_4, \tag{4.40}$$

where

$$N_1 = \frac{1}{4}(1 - X_1)(1 - X_2)\,; \quad N_2 = \frac{1}{4}(1 + X_1)(1 - X_2) \tag{4.41a}$$

$$N_3 = \frac{1}{4}(1 + X_1)(1 + X_2)\,; \quad N_4 = \frac{1}{4}(1 - X_1)(1 + X_2). \tag{4.41b}$$

(a) For $\boldsymbol{X} = (0,0)^{\mathrm{T}}$ and $\boldsymbol{x}_1 = (4,2)^{\mathrm{T}}$, $\boldsymbol{x}_2 = (8,4)^{\mathrm{T}}$, $\boldsymbol{x}_3 = (6,8)^{\mathrm{T}}$, $\boldsymbol{x}_4 = (4,6)^{\mathrm{T}}$ find the deformation gradient $\boldsymbol{F}$ and Green's strain $\boldsymbol{E}$. (b) Find the principal stretches λ_1 and λ_2 and the corresponding principal material and spatial unit vectors $\boldsymbol{N}_1$, $\boldsymbol{N}_2$, and $\boldsymbol{n}_1$, $\boldsymbol{n}_2$ respectively. (c) Using the above calculations show that Green's strain tensor can be expressed in terms of the principal material directions as

$$\boldsymbol{E} = \frac{1}{2}\big[(\lambda_1^2 - 1)\boldsymbol{N}_1 \otimes \boldsymbol{N}_1 + (\lambda_2^2 - 1)\boldsymbol{N}_2 \otimes \boldsymbol{N}_2\big]. \tag{4.42}$$

Solution

Substituting the nodal spatial coordinates into Equation (4.40) and differentiating with respect to the material coordinates gives

$$\frac{\partial x_1}{\partial X_1} = -\frac{1}{4}(1-X_2)4 + \frac{1}{4}(1-X_2)8 + \frac{1}{4}(1+X_2)6 - \frac{1}{4}(1+X_2)4, \tag{4.43a}$$

$$\frac{\partial x_2}{\partial X_1} = -\frac{1}{4}(1-X_2)2 + \frac{1}{4}(1-X_2)4 + \frac{1}{4}(1+X_2)8 - \frac{1}{4}(1+X_2)6, \tag{4.43b}$$

$$\frac{\partial x_1}{\partial X_2} = -\frac{1}{4}(1-X_1)4 - \frac{1}{4}(1-X_1)8 + \frac{1}{4}(1+X_1)6 + \frac{1}{4}(1-X_1)4, \tag{4.43c}$$

$$\frac{\partial x_2}{\partial X_2} = -\frac{1}{4}(1-X_1)2 - \frac{1}{4}(1-X_1)4 + \frac{1}{4}(1+X_1)8 + \frac{1}{4}(1-X_1)6. \tag{4.43d}$$

Substituting the material coordinates $\boldsymbol{X} = (0,0)^{\mathrm{T}}$ gives

$$\boldsymbol{F} = \begin{bmatrix} \partial x_1/\partial X_1 & \partial x_1/\partial X_2 \\ \partial x_2/\partial X_1 & \partial x_2/\partial X_2. \end{bmatrix} = \begin{bmatrix} 1.5 & -0.5 \\ 1.0 & 2.0 \end{bmatrix}. \tag{4.44}$$

Green's strain tensor $\boldsymbol{E}$ is easily found from Equation (4.4) as

$$\boldsymbol{E} = \frac{1}{2}(\boldsymbol{F}^{\mathrm{T}}\boldsymbol{F} - \boldsymbol{I}) = \begin{bmatrix} 1.125 & 0.625 \\ 0.625 & 1.625 \end{bmatrix}. \tag{4.45}$$

From Equation (4.6) the stretches λ_1 and λ_2 are the square root of the eigenvalues of $\boldsymbol{C}$ and the principal material vectors $\boldsymbol{N}_1$ and $\boldsymbol{N}_2$ are the corresponding normalized eigenvectors, yielding $\lambda_1 = 1.5504$ and $\lambda_2 = 2.2575$ and

$$\boldsymbol{N}_1 = \begin{bmatrix} -0.8281 \\ 0.5606 \end{bmatrix}; \quad \boldsymbol{N}_2 = \begin{bmatrix} 0.5606 \\ 0.8281 \end{bmatrix}. \tag{4.46}$$

The rotation tensor is given by Equation (4.5) as $\boldsymbol{R} = \boldsymbol{F}\boldsymbol{U}^{-1}$ where $\boldsymbol{U}$ is given by Equation (4.7), hence

$$\boldsymbol{U} = 1.5504 \begin{bmatrix} -0.8281 \\ 0.5606 \end{bmatrix} [-0.8281, 0.5606] + 2.2575 \begin{bmatrix} 0.5606 \\ 0.8281 \end{bmatrix} [0.5606, 0.8281] \tag{4.47a}$$

$$= \begin{bmatrix} 1.7726 & 0.3283 \\ 0.3283 & 2.0352 \end{bmatrix}. \tag{4.47b}$$

Hence

$$\boldsymbol{R} = \begin{bmatrix} 0.9191 & -0.3939 \\ 0.3939 & 0.9191 \end{bmatrix}. \tag{4.48}$$

The spatial unit vectors are given by $\boldsymbol{n}_\alpha = \boldsymbol{R}\boldsymbol{N}_\alpha$, $\alpha = 1, 2$ as

$$\boldsymbol{n}_1 = \begin{bmatrix} -0.9820 \\ 0.1891 \end{bmatrix}; \quad \boldsymbol{n}_2 = \begin{bmatrix} 0.1891 \\ 0.9820 \end{bmatrix}. \tag{4.49}$$

Figure 4.2 shows the principal unit vectors before and the corresponding rotated and stretch vectors after deformation for the material point $\boldsymbol{X} = (0, 0)^{\mathrm{T}}$. The deformation information for other material points can readily be explored using the program "polar_decomposition.m."

Using a similar calculation to that given in Equation (4.47) together with Equations (4.4), (4.6) it is easy to demonstrate that Green's strain $\boldsymbol{E}$ can be expressed in principal directions as given by Equation (4.42).

EXAMPLE 4.4

The finite deformation of a two-dimensional continuum from initial position $\boldsymbol{X} = (X_1, X_2)^{\mathrm{T}}$ to a final configuration $\boldsymbol{x} = (x_1, x_2)^{\mathrm{T}}$ is given as

$$x_1 = 4 - 2X_1 - X_2\,; \quad x_2 = 2 + \frac{3}{2}X_1 - \frac{1}{2}X_2. \tag{4.50}$$

(a) Calculate the deformation gradient tensor $\boldsymbol{F}$, the right Cauchy–Green strain tensor $\boldsymbol{C}$ and the left Cauchy–Green strain tensor $\boldsymbol{b}$.

(b) Calculate the stretch undergone by a unit material vector $\boldsymbol{a}_0 = (\frac{3}{5}, \frac{4}{5})^{\mathrm{T}}$ as a result of the deformation.

(c) A pair of orthonormal material vectors $\boldsymbol{b}_0 = (1, 0)^{\mathrm{T}}$ and $\boldsymbol{c}_0 = (0, 1)^{\mathrm{T}}$ are subjected to the deformation process. Calculate the new angle formed between these two vectors after deformation.
(d) Demonstrate whether or not the deformation is isochoric.
(e) Calculate the principal stretches λ_α and the principal material directions $\boldsymbol{N}_\alpha$, with $\alpha = 1, 2$.
(f) Calculate the principal spatial directions $\boldsymbol{n}_\alpha$, with $\alpha = 1, 2$.

Solution

(a) The deformation gradient, the right Cauchy–Green strain tensor, and the left Cauchy–Green strain tensor are easily found from Equation (4.50) as

$$\boldsymbol{F} = \begin{bmatrix} \partial x_1/\partial X_1 & \partial x_1/\partial X_2 \\ \partial x_2/\partial X_1 & \partial x_2/\partial X_2. \end{bmatrix} = \begin{bmatrix} -2 & -1 \\ 3/2 & -1/2 \end{bmatrix}. \tag{4.51}$$

The right Cauchy–Green tensor $\boldsymbol{C} = \boldsymbol{F}^{\mathrm{T}}\boldsymbol{F}$ and the left Cauchy–Green tensor $\boldsymbol{b} = \boldsymbol{F}\boldsymbol{F}^{\mathrm{T}}$, yielding

$$\begin{aligned} \boldsymbol{C} &= \begin{bmatrix} -2 & 3/2 \\ -1 & -1/2 \end{bmatrix} \begin{bmatrix} -2 & -1 \\ 3/2 & -1/2 \end{bmatrix} \\ &= \begin{bmatrix} 6.25 & 1.25 \\ 1.25 & 1.25 \end{bmatrix}, \end{aligned} \tag{4.52}$$

and

$$\boldsymbol{b} = \begin{bmatrix} 5 & -2.5 \\ -2.5 & 2.5 \end{bmatrix}. \tag{4.53}$$

(b) Recall textbook **Remark 4.3**: The general nature of the scalar product as a measure of deformation can be clarified by taking $d\boldsymbol{X}_2$ and $d\boldsymbol{X}_1$ equal to $d\boldsymbol{X}$ and consequently $d\boldsymbol{x}_1 = d\boldsymbol{x}_2 = d\boldsymbol{x}$. This enables initial (material) and current (spatial) elemental lengths squared to be determined as

$$dL^2 = d\boldsymbol{X} \cdot d\boldsymbol{X}; \qquad dl^2 = d\boldsymbol{x} \cdot d\boldsymbol{x}. \tag{4.54}$$

The change in the squared lengths that occurs as the body deforms from the initial to the current configuration can now be written in terms of the elemental material vector $d\boldsymbol{X}$ as

$$\frac{1}{2}(dl^2 - dL^2) = d\boldsymbol{X} \cdot \boldsymbol{E}\, d\boldsymbol{X}, \tag{4.55}$$

which, upon division by dL^2, gives the scalar Green's strain as

$$\frac{dl^2 - dL^2}{2\, dL^2} = \frac{d\boldsymbol{X}}{dL} \cdot \boldsymbol{E}\, \frac{d\boldsymbol{X}}{dL}, \tag{4.56}$$

where $d\boldsymbol{X}/dL$ is a unit material vector $\boldsymbol{N}$ in the direction of $d\boldsymbol{X}$; hence, finally

$$\frac{1}{2}\left(\frac{dl^2 - dL^2}{dL^2}\right) = \boldsymbol{N} \cdot \boldsymbol{E}\boldsymbol{N}. \tag{4.57}$$

From Equations (4.9) and (4.4), Equation (4.57) can be rewritten in terms of the stretch of the unit vector $\boldsymbol{N}$ as

$$\begin{aligned} \lambda^2 - 1 &= \boldsymbol{N} \cdot (\boldsymbol{C} - \boldsymbol{I})\boldsymbol{N} \\ &= \boldsymbol{N} \cdot \boldsymbol{C}\boldsymbol{N} - 1. \end{aligned} \tag{4.58}$$

Substituting $\boldsymbol{a}_0$ for $\boldsymbol{N}$ gives

$$\lambda_{\boldsymbol{a}_0}^2 = \boldsymbol{a}_0 \cdot \boldsymbol{C}\boldsymbol{a}_0. \tag{4.59}$$

Substituting $\boldsymbol{a}_0 = [\,\frac{3}{5}, \frac{4}{5}\,]^T$ and using Equation (4.52) gives

$$\lambda_{\boldsymbol{a}_0}^2 = \left[\frac{3}{5}, \frac{4}{5}\right] \begin{bmatrix} 6.25 & 1.25 \\ 1.25 & 1.25 \end{bmatrix} \begin{bmatrix} 3/5 \\ 4/5 \end{bmatrix} = 4.25\ ;\ \lambda_{\boldsymbol{a}_0} = 2.061. \tag{4.60}$$

(c) The spatial vectors corresponding to material vectors $\boldsymbol{b}_0$ and $\boldsymbol{c}_0$ are found using the deformation gradient $\boldsymbol{F}$ as

$$\boldsymbol{b} = \boldsymbol{F}\boldsymbol{b}_0\ ;\ \boldsymbol{c} = \boldsymbol{F}\boldsymbol{c}_0. \tag{4.61}$$

Using the cosine rule enables the angle θ_{bc} between spatial vectors $\boldsymbol{b}$ and $\boldsymbol{c}$ to be found as

$$\theta_{bc} = \cos^{-1}\left[\frac{\boldsymbol{b} \cdot \boldsymbol{c}}{\|\boldsymbol{b}\|\, \|\boldsymbol{c}\|}\right] = \cos^{-1}\left[\frac{\boldsymbol{b}_0 \cdot \boldsymbol{C}\boldsymbol{c}_0}{(\boldsymbol{b}_0 \cdot \boldsymbol{C}\boldsymbol{b}_0)^{\frac{1}{2}}\, (\boldsymbol{c}_0 \cdot \boldsymbol{C}\boldsymbol{c}_0)^{\frac{1}{2}}}\right]. \tag{4.62}$$

It is now a simple matter of substitution to find θ_{bc} as

$$\boldsymbol{b}_0 \cdot \boldsymbol{C}\boldsymbol{c}_0 = [1,0] \begin{bmatrix} 6.25 & 1.25 \\ 1.25 & 1.25 \end{bmatrix} \begin{bmatrix} 0 \\ 1 \end{bmatrix} = 1.25, \tag{4.63a}$$

$$\boldsymbol{b}_0 \cdot \boldsymbol{C}\boldsymbol{b}_0 = 6.25 \; ; \; \boldsymbol{c}_0 \cdot \boldsymbol{C}\boldsymbol{c}_0 = 1.25, \tag{4.63b}$$

$$\theta_{bc} = \cos^{-1} \left[\frac{1.25}{(6.25 \times 1.25)^{\frac{1}{2}}} \right] = 63.43^{\circ}. \tag{4.63c}$$

(d) To determine whether the deformation is isochoric requires checking the determinant of the deformation gradient as

$$J = \det(\boldsymbol{F}) = \left(1 + \frac{3}{2}\right) = 2.5, \tag{4.64}$$

which being greater than unit indicates the deformation is not isochoric.

(e) The stretches are found as in the previous example by calculating the square root of the eigenvalues of the right Cauchy–Green tensor $\boldsymbol{C}$ to give $\lambda_1 = 2.5583$ and $\lambda_2 = 0.9772$. The corresponding normalized eigenvectors $\boldsymbol{N}_1$ and $\boldsymbol{N}_2$ yield the principal material directions

$$\boldsymbol{N}_1 = \begin{bmatrix} 0.9732 \\ 0.2298 \end{bmatrix} \; ; \quad \boldsymbol{N}_2 = \begin{bmatrix} -0.2298 \\ 0.9732 \end{bmatrix}. \tag{4.65}$$

(f) As an alternative to finding the spatial principal vectors as shown in Equation (4.49), $\boldsymbol{n}_1$ and $\boldsymbol{n}_2$ can be found by normalizing the push forward of the material principal vectors $\boldsymbol{N}_1$ and $\boldsymbol{N}_2$ to give

$$\boldsymbol{n}_1 = \frac{\boldsymbol{F}\boldsymbol{N}_1}{\|\boldsymbol{F}\boldsymbol{N}_1\|} = \begin{bmatrix} -0.8507 \\ 0.5257 \end{bmatrix}, \tag{4.66a}$$

$$\boldsymbol{n}_2 = \frac{\boldsymbol{F}\boldsymbol{N}_2}{\|\boldsymbol{F}\boldsymbol{N}_2\|} = \begin{bmatrix} -0.5257 \\ -0.8507 \end{bmatrix}. \tag{4.66b}$$

EXAMPLE 4.5: Textbook Exercise 4.2

(a) If the deformation gradients at times t and $t + \Delta t$ are $\boldsymbol{F}_t$ and $\boldsymbol{F}_{t+\Delta t}$ respectively, show that the deformation gradient $\Delta\boldsymbol{F}$ relating the incremental motion from configuration at t to $t + \Delta t$ is $\Delta\boldsymbol{F} = \boldsymbol{F}_{t+\Delta t}\boldsymbol{F}_t^{-1}$.

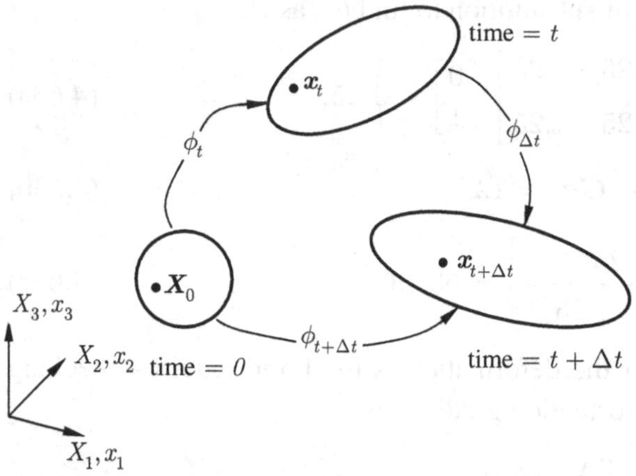

FIGURE 4.3 General motion of a deformable body

(b) Using the deformation given in textbook Example 4.5 with $\boldsymbol{X} = (0, 0)^{\mathrm{T}}$, $t = 1$, $\Delta t = 1$, show that $\Delta\boldsymbol{F} = \boldsymbol{F}_{t+\Delta t}\boldsymbol{F}_t^{-1}$ is correct by pushing forward the initial vector $\boldsymbol{G} = [1, 1]^{\mathrm{T}}$ to vectors $\boldsymbol{g}_t$ and $\boldsymbol{g}_{t+\Delta t}$ at times t and $t + \Delta t$ respectively and checking that $\boldsymbol{g}_{t+\Delta t} = \Delta\boldsymbol{F}\boldsymbol{g}_t$.
(c) [Addition to textbook] Using the deformation given in textbook Example 4.5 with $\boldsymbol{X} = (0, 0)^{\mathrm{T}}$ and $t = 1$, calculate the velocity gradient $\boldsymbol{l}$ and the rate of deformation $\boldsymbol{d}$.

Solution

(a) Deformation gradients relating to the various configurations shown in Figure 4.3 can be defined as

$$\boldsymbol{F}_t = \frac{\partial \boldsymbol{x}_t}{\partial \boldsymbol{X}}, \tag{4.67a}$$

$$\boldsymbol{F}_{t+\Delta t} = \frac{\partial \boldsymbol{x}_{t+\Delta t}}{\partial \boldsymbol{X}}, \tag{4.67b}$$

$$\Delta\boldsymbol{F} = \frac{\partial \boldsymbol{x}_{t+\Delta t}}{\partial \boldsymbol{x}_t}. \tag{4.67c}$$

From Equation (4.3), which relates an elemental material vector $d\boldsymbol{X}$ to the corresponding spatial vector $d\boldsymbol{x}$, relationships can be established between elemental material and spatial vectors for the configurations shown in

Figure 4.3:

$$d\boldsymbol{x}_t = \boldsymbol{F}_t d\boldsymbol{X}, \tag{4.68a}$$

$$d\boldsymbol{x}_{t+\Delta t} = \boldsymbol{F}_{t+\Delta t} d\boldsymbol{X}, \tag{4.68b}$$

$$d\boldsymbol{x}_{t+\Delta t} = \Delta \boldsymbol{F} d\boldsymbol{x}_t. \tag{4.68c}$$

Substituting Equation (4.68a) into (4.68c) gives

$$d\boldsymbol{x}_{t+\Delta t} = \Delta \boldsymbol{F} \boldsymbol{F}_t d\boldsymbol{X}. \tag{4.69}$$

Substituting Equation (4.68b) into (4.69) yields

$$\boldsymbol{F}_{t+\Delta t} d\boldsymbol{X} = \Delta \boldsymbol{F} \boldsymbol{F}_t d\boldsymbol{X}. \tag{4.70}$$

Consequently,

$$\Delta \boldsymbol{F} = \boldsymbol{F}_{t+\Delta t} \boldsymbol{F}_t^{-1}. \tag{4.71}$$

Alternatively the chain rule can be invoked to give a direct solution as

$$\begin{aligned}
\boldsymbol{F}_{t+\Delta t} &= \frac{\partial \boldsymbol{x}_{t+\Delta t}}{\partial \boldsymbol{X}} \\
&= \frac{\partial \boldsymbol{x}_{t+\Delta t}}{\partial \boldsymbol{x}_t} \frac{\partial \boldsymbol{x}_t}{\partial \boldsymbol{X}} \\
&= \Delta \boldsymbol{F} \boldsymbol{F}_t,
\end{aligned} \tag{4.72}$$

which, rearranged, gives Equation (4.71).

(b) The deformation is given by

$$x_1 = \tfrac{1}{4}(4X_1 + (9 - 3X_1 - 5X_2 - X_1X_2)t)$$

$$x_2 = \tfrac{1}{4}(4X_2 + (16 + 8X_1)t).$$

From this, the components of the deformation gradient are found to be

$$F_{11} = \frac{\partial x_1}{\partial X_1} = \frac{1}{4}(4 + (-3 - X_2)t) = \frac{4 - (3 + X_2)t}{4} \tag{4.73a}$$

$$F_{12} = \frac{\partial x_1}{\partial X_2} = \frac{1}{4}((-5 - X_1)t) = \frac{-(5 + X_1)t}{4} \tag{4.73b}$$

$$F_{21} = \frac{\partial x_2}{\partial X_1} = \frac{1}{4}(8t) = 2t \tag{4.73c}$$

$$F_{22} = \frac{\partial x_2}{\partial X_2} = \frac{1}{4}(4) = 1. \tag{4.73d}$$

For $\boldsymbol{X} = [0,0]^{\mathrm{T}}$ this yields

$$\boldsymbol{F} = \begin{bmatrix} 1-\frac{3}{4}t & -\frac{5t}{4} \\ 2t & 1 \end{bmatrix}. \tag{4.74}$$

Thus for $t = 1$ and $t + \Delta t = 2$,

$$\boldsymbol{F}_t = \begin{bmatrix} \frac{1}{4} & -\frac{5}{4} \\ 2 & 1 \end{bmatrix}; \quad \boldsymbol{F}_{t+\Delta t} = \begin{bmatrix} -\frac{1}{2} & -\frac{5}{2} \\ 4 & 1 \end{bmatrix}, \tag{4.75}$$

which enables $\Delta\boldsymbol{F}$ to be calculated as

$$\begin{aligned} \Delta\boldsymbol{F} = \boldsymbol{F}_{t+\Delta t}\boldsymbol{F}_t^{-1} &= \begin{bmatrix} -\frac{1}{2} & -\frac{5}{2} \\ 4 & 1 \end{bmatrix} \begin{bmatrix} \frac{4}{11} & -\frac{5}{11} \\ -\frac{8}{11} & \frac{1}{11} \end{bmatrix} \\ &= \begin{bmatrix} \frac{18}{11} & -\frac{5}{11} \\ -\frac{8}{11} & \frac{21}{11} \end{bmatrix}. \end{aligned} \tag{4.76}$$

The push forward of the material vector $\boldsymbol{G} = [1,1]^{\mathrm{T}}$ at times t and $t + \Delta t$ becomes

$$\boldsymbol{g}_t = \boldsymbol{F}_t\boldsymbol{G} = \begin{bmatrix} \frac{1}{4} & -\frac{5}{4} \\ 2 & 1 \end{bmatrix} \begin{bmatrix} 1 \\ 1 \end{bmatrix} = \begin{bmatrix} -1 \\ 3 \end{bmatrix}, \tag{4.77a}$$

$$\boldsymbol{g}_{t+\Delta t} = \boldsymbol{F}_{t+\Delta t}\boldsymbol{G} = \begin{bmatrix} -\frac{1}{2} & -\frac{5}{2} \\ 4 & 1 \end{bmatrix} \begin{bmatrix} 1 \\ 1 \end{bmatrix} = \begin{bmatrix} -3 \\ 5 \end{bmatrix}. \tag{4.77b}$$

Finally $\boldsymbol{g}_{t+\Delta t}$ can be calculated using Equation (4.68) to give

$$\boldsymbol{g}_{t+\Delta t} = \Delta\boldsymbol{F}\boldsymbol{g}_t = \begin{bmatrix} \frac{18}{11} & -\frac{5}{11} \\ -\frac{8}{11} & \frac{21}{11} \end{bmatrix} \begin{bmatrix} -1 \\ 3 \end{bmatrix} = \begin{bmatrix} -3 \\ 5 \end{bmatrix}. \tag{4.78}$$

(c) The velocity gradient tensor is given by Equation (4.20) as $\boldsymbol{l} = \dot{\boldsymbol{F}}\boldsymbol{F}^{-1}$ where the inverse of the deformation gradient tensor at $t = 1$ given in Equation (4.75) is

$$\boldsymbol{F}_t^{-1} = \begin{bmatrix} \frac{4}{11} & \frac{5}{11} \\ -\frac{8}{11} & \frac{1}{11} \end{bmatrix}. \tag{4.79}$$

The time derivative of the deformation gradient at $t = 1$ is obtained from Equation (4.74) as

$$\dot{\boldsymbol{F}}_t = \begin{bmatrix} -\frac{3}{4} & -\frac{5}{4} \\ 2 & 0 \end{bmatrix}. \tag{4.80}$$

The velocity gradient is now easily found as

$$\boldsymbol{l} = \dot{\boldsymbol{F}}_t \boldsymbol{F}_t^{-1} = \begin{bmatrix} \frac{7}{11} & -\frac{5}{11} \\ \frac{8}{11} & \frac{10}{11} \end{bmatrix}, \tag{4.81}$$

and from Equation (4.23) the rate of deformation tensor is calculated as

$$\begin{aligned} \boldsymbol{d} &= \tfrac{1}{2}(\boldsymbol{l} + \boldsymbol{l}^T) \\ &= \tfrac{1}{2}\left(\begin{bmatrix} \frac{7}{11} & -\frac{5}{11} \\ \frac{8}{11} & \frac{10}{11} \end{bmatrix} + \begin{bmatrix} \frac{7}{11} & \frac{8}{11} \\ -\frac{5}{11} & \frac{10}{11} \end{bmatrix} \right) \\ &= \begin{bmatrix} \frac{14}{22} & \frac{3}{22} \\ \frac{3}{22} & \frac{20}{22} \end{bmatrix}. \end{aligned} \tag{4.82}$$

EXAMPLE 4.6: Textbook Exercise 4.3

Using Equation (4.14) prove that the area ratio can be expressed alternatively as

$$\frac{da}{dA} = J\sqrt{\boldsymbol{N} \cdot \boldsymbol{C}^{-1}\boldsymbol{N}}.$$

Solution

Equation (4.14) can be expanded to give

$$da\,\boldsymbol{n} = J\boldsymbol{F}^{-\mathrm{T}} dA\,\boldsymbol{N}. \tag{4.83}$$

Squaring both sides yields

$$da^2 \boldsymbol{n} \cdot \boldsymbol{n} = J^2 dA^2 (\boldsymbol{F}^{-\mathrm{T}}\boldsymbol{N})^{\mathrm{T}} (\boldsymbol{F}^{-\mathrm{T}}\boldsymbol{N}). \tag{4.84}$$

Noting that $\boldsymbol{n}\cdot\boldsymbol{n} = 1$ and the right Cauchy–Green tensor $\boldsymbol{C} = \boldsymbol{F}^{\mathrm{T}}\boldsymbol{F}$ allows da^2 to be extracted as

$$\begin{aligned} da^2 &= J^2 dA^2 \boldsymbol{N}^{\mathrm{T}}(\boldsymbol{F}^{-1}\boldsymbol{F}^{-\mathrm{T}})\boldsymbol{N} \\ &= J^2 dA^2 \boldsymbol{N}\cdot\boldsymbol{C}^{-1}\boldsymbol{N}, \end{aligned} \tag{4.85}$$

gives

$$\frac{da}{dA} = J\sqrt{\boldsymbol{N}\cdot\boldsymbol{C}^{-1}\boldsymbol{N}}. \tag{4.86}$$

EXAMPLE 4.7: Textbook Exercise 4.4

Consider the planar (1–2) deformation for which the deformation gradient is

$$\boldsymbol{F} = \begin{bmatrix} F_{11} & F_{12} & 0 \\ F_{21} & F_{22} & 0 \\ 0 & 0 & \lambda_3 \end{bmatrix},$$

where λ_3 is the stretch in the thickness direction normal to the (1–2) plane. If dA and da are the elemental areas in the (1–2) plane and H and h the thicknesses before and after deformation respectively, show that

$$\frac{da}{dA} = j \quad \text{and} \quad h = H\frac{J}{j},$$

where $j = \det(F_{kl}),\ k, l = 1, 2$.

Solution

From Equation (4.12) the elemental volume ratio is given by

$$dv = \det \boldsymbol{F} dV\ ; \det \boldsymbol{F} = j\lambda_3 \quad \lambda_3 = \frac{h}{H}, \tag{4.87}$$

which can be rewritten in terms of h and H as

$$h\, da = H \det \boldsymbol{F} dA. \tag{4.88}$$

Substituting from Equation (4.87) gives

$$h\, da = H\, j\frac{h}{H} dA, \tag{4.89}$$

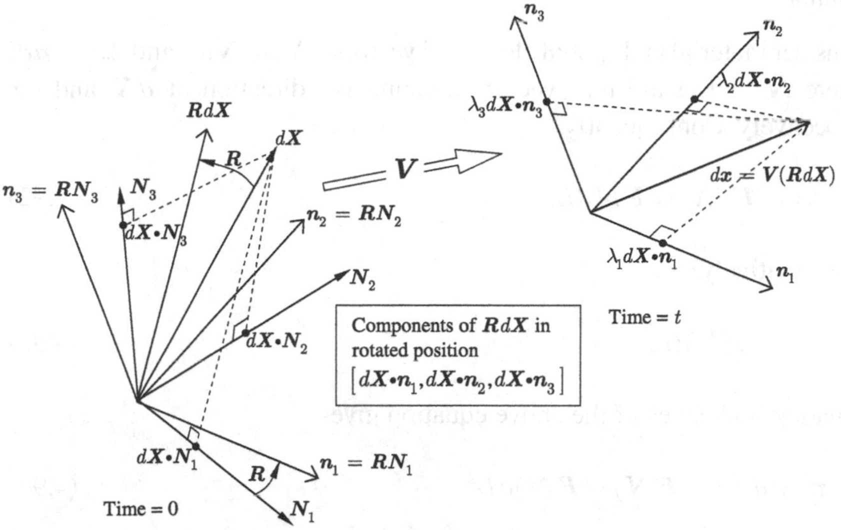

FIGURE 4.4 Interpretation of $dx = VRdX$

hence,

$$\frac{da}{dA} = j, \tag{4.90}$$

and from Equation (4.87)

$$h = H\,\frac{\det \boldsymbol{F}}{j} = H\,\frac{J}{j}. \tag{4.91}$$

EXAMPLE 4.8: Textbook Exercise 4.5

Using textbook Figure 4.4 as a guide, draw a similar diagram that interprets the polar decomposition Equation (4.8) $dx = \boldsymbol{V}(\boldsymbol{R}d\boldsymbol{X})$.

Solution

See Figure 4.4.

EXAMPLE 4.9: Textbook Exercise 4.6

Show that the condition for an elemental material vector $d\boldsymbol{X} = \boldsymbol{N}dL$ to exhibit zero extension is $\boldsymbol{N} \cdot \boldsymbol{C}\boldsymbol{N} = 1$, where $\boldsymbol{C} = \boldsymbol{F}^{\mathrm{T}}\boldsymbol{F}$.

Solution

Consider material and spatial elemental vectors $d\boldsymbol{X} = \boldsymbol{N}dL$ and $d\boldsymbol{x} = \boldsymbol{n}dl$ where $\boldsymbol{N}$ and $\boldsymbol{n}$ are unit vectors defining the direction of $d\boldsymbol{X}$ and $d\boldsymbol{x}$ respectively. Consequently,

$$d\boldsymbol{x} = \boldsymbol{F}d\boldsymbol{X} = \boldsymbol{F}\boldsymbol{N}dL, \tag{4.92}$$

or alternatively,

$$\boldsymbol{n}dl = \boldsymbol{F}\boldsymbol{N}dL. \tag{4.93}$$

Squaring both sides of the above equation gives

$$\boldsymbol{n}\cdot\boldsymbol{n}dl^2 = (\boldsymbol{F}\boldsymbol{N})\cdot(\boldsymbol{F}\boldsymbol{N})dL^2. \tag{4.94}$$

Hence,

$$\left(\frac{dl}{dL}\right)^2 = \boldsymbol{N}^{\mathrm{T}}(\boldsymbol{F}^{\mathrm{T}}\boldsymbol{F})\boldsymbol{N} = \boldsymbol{N}\cdot\boldsymbol{C}\boldsymbol{N}, \tag{4.95}$$

implying that for $dl = dL$, we have $\boldsymbol{N}\cdot\boldsymbol{C}\boldsymbol{N} = 1$.

EXAMPLE 4.10: Textbook Exercise 4.7

Prove Equation (4.11), that is,

$$\boldsymbol{F}^{-\mathrm{T}}\boldsymbol{N}_\alpha = \frac{1}{\lambda_\alpha}\boldsymbol{n}_\alpha.$$

Solution

It is instructive to prove that $\boldsymbol{U}\boldsymbol{N}_\alpha = \lambda_\alpha\boldsymbol{N}_\alpha$ as given at the bottom of textbook page 107 as this is relevant to proving Equation (4.10). Recall from Equation (4.7) that

$$\boldsymbol{U} = \sum_{\beta=1}^{3}\lambda_\beta\boldsymbol{N}_\beta\otimes\boldsymbol{N}_\beta, \tag{4.96}$$

hence

$$\begin{aligned} \boldsymbol{U}\boldsymbol{N}_\alpha &= \left(\sum_{\beta=1}^{3} \lambda_\beta \boldsymbol{N}_\beta \otimes \boldsymbol{N}_\beta\right)\boldsymbol{N}_\alpha \\ &= \sum_{\beta=1}^{3} \lambda_\beta (\boldsymbol{N}_\alpha \cdot \boldsymbol{N}_\beta)\boldsymbol{N}_\beta \\ &= \lambda_\alpha \boldsymbol{N}_\alpha. \end{aligned} \tag{4.97}$$

In a similar manner and in preparation for solving this Exercise, it can be shown using Equation (4.6) that

$$\begin{aligned} \boldsymbol{C}^{-1}\boldsymbol{N}_\alpha &= \left(\sum_{\beta=1}^{3} \frac{1}{\lambda_\beta^2} \boldsymbol{N}_\beta \otimes \boldsymbol{N}_\beta\right)\boldsymbol{N}_\alpha \\ &= \frac{1}{\lambda_\alpha^2}\boldsymbol{N}_\alpha. \end{aligned} \tag{4.98}$$

Multiplying $\boldsymbol{F}\boldsymbol{N}_\alpha = \boldsymbol{R}\boldsymbol{U}\boldsymbol{N}_\alpha$ through by $\boldsymbol{F}^{-\mathrm{T}}\boldsymbol{F}^{-1}$ and noting that $\boldsymbol{U}\boldsymbol{N}_\alpha = \lambda_\alpha \boldsymbol{N}_\alpha$ gives

$$\begin{aligned} \boldsymbol{F}^{-\mathrm{T}}\boldsymbol{N}_\alpha &= \boldsymbol{F}^{-\mathrm{T}}\boldsymbol{F}^{-1}\boldsymbol{R}\boldsymbol{U}\boldsymbol{N}_\alpha \\ &= \lambda_\alpha(\boldsymbol{F}^{-\mathrm{T}})(\boldsymbol{F}^{-1})\boldsymbol{R}\boldsymbol{N}_\alpha. \end{aligned} \tag{4.99}$$

Recalling that $\boldsymbol{R}^{-1} = \boldsymbol{R}^{\mathrm{T}}$ and $\boldsymbol{U}^{\mathrm{T}} = \boldsymbol{U}$ yields

$$\begin{aligned} \boldsymbol{F}^{-\mathrm{T}}\boldsymbol{N}_\alpha &= \lambda_\alpha(\boldsymbol{R}^{-\mathrm{T}}\boldsymbol{U}^{-\mathrm{T}})(\boldsymbol{U}^{-1}\boldsymbol{R}^{-1})\boldsymbol{R}\boldsymbol{N}_\alpha \\ &= \lambda_\alpha \boldsymbol{R}\boldsymbol{U}^{-1}\boldsymbol{U}^{-1}\boldsymbol{N}_\alpha \\ &= \lambda_\alpha \boldsymbol{R}\boldsymbol{C}^{-1}\boldsymbol{N}_\alpha \\ &= \lambda_\alpha \boldsymbol{R}\frac{1}{\lambda_\alpha^2}\boldsymbol{N}_\alpha \\ &= \frac{1}{\lambda_\alpha}\boldsymbol{R}\boldsymbol{N}_\alpha \\ &= \frac{1}{\lambda_\alpha}\boldsymbol{n}_\alpha. \end{aligned} \tag{4.100}$$

EXAMPLE 4.11: Textbook Exercise 4.8

The motion of a body, at time t, is given by

$$\boldsymbol{x} = \boldsymbol{F}(t)\boldsymbol{X}; \qquad \boldsymbol{F}(t) = \begin{bmatrix} 1 & t & t^2 \\ t^2 & 1 & t \\ t & t^2 & 1 \end{bmatrix}; \tag{4.101a}$$

$$\boldsymbol{F}^{-1}(t) = \frac{1}{(t^3-1)} \begin{bmatrix} -1 & t & 0 \\ 0 & -1 & t \\ t & 0 & -1 \end{bmatrix}. \tag{4.101b}$$

Find the velocity of the particle, (a) initially at $\boldsymbol{X} = (1,1,1)^{\mathrm{T}}$ at time $t=0$; and (b) currently at $\boldsymbol{x} = (1,1,1)^{\mathrm{T}}$ at time $t=2$. Using $J = dv/dV$ show that at time $t=1$ the motion is not realistic.

Solution

From Equations (4.19) and (4.101) the velocity can be found to be

$$\boldsymbol{v}(\boldsymbol{X},t) = \frac{\partial \boldsymbol{\phi}(\boldsymbol{X},t)}{\partial t} = \frac{\partial \boldsymbol{F}(t)}{\partial t}\boldsymbol{X} = \begin{bmatrix} 0 & 1 & 2t \\ 2t & 0 & 1 \\ 1 & 2t & 0 \end{bmatrix} \boldsymbol{X}. \tag{4.102}$$

Consequently at time $t=0$ and $\boldsymbol{X} = (1,1,1)^{\mathrm{T}}$ the velocity $\boldsymbol{v} = (1,1,1)^{\mathrm{T}}$. The material coordinates $\boldsymbol{X}$ can be written in terms of the spatial coordinates $\boldsymbol{x}$ as $\boldsymbol{X} = \boldsymbol{F}^{-1}(t)\boldsymbol{x}$; thus for time $t=2$ and $\boldsymbol{x} = (1,1,1)^{\mathrm{T}}$, the material coordinates are calculated as

$$\boldsymbol{X} = \frac{1}{(t^3-1)} \begin{bmatrix} -1 & t & 0 \\ 0 & -1 & t \\ t & 0 & -1 \end{bmatrix}_{t=2} \begin{bmatrix} 1 \\ 1 \\ 1 \end{bmatrix} = \frac{1}{7} \begin{bmatrix} 1 \\ 1 \\ 1 \end{bmatrix}. \tag{4.103}$$

Using the above material coordinates together with Equation (4.102), the velocity at $t=2$ is simply found to be

$$\boldsymbol{v} = \begin{bmatrix} 0 & 1 & 4 \\ 4 & 0 & 1 \\ 1 & 4 & 0 \end{bmatrix} \frac{1}{7} \begin{bmatrix} 1 \\ 1 \\ 1 \end{bmatrix} = \frac{5}{7} \begin{bmatrix} 1 \\ 1 \\ 1 \end{bmatrix}. \tag{4.104}$$

Finally, $\det \boldsymbol{F} = t^3 - 1$ hence at $t = 1$ $\det \boldsymbol{F} = 0$; that is, the material has vanished!

EXAMPLE 4.12: Textbook Exercise 4.9

For a pure expansion the deformation gradient is $\boldsymbol{F} = \alpha \boldsymbol{I}$, where α is a scalar (function of time). Show that the rate of deformation is

$$\boldsymbol{d} = \frac{\dot{\alpha}}{\alpha}\boldsymbol{I}.$$

Solution

The material time derivative of $\boldsymbol{F}$ is easily determined as $\dot{\boldsymbol{F}} = \dot{\alpha}\boldsymbol{I}$. From Equation (4.21) the material time derivative of Green's strain is found to be

$$\begin{aligned}\dot{\boldsymbol{E}} &= \tfrac{1}{2}(\dot{\boldsymbol{F}}^{\mathrm{T}}\boldsymbol{F} + \boldsymbol{F}^{\mathrm{T}}\dot{\boldsymbol{F}})\\ &= \tfrac{1}{2}\dot{\alpha}(\boldsymbol{F} + \boldsymbol{F}^{\mathrm{T}})\\ &= \tfrac{1}{2}\dot{\alpha}(\alpha\boldsymbol{I} + \alpha\boldsymbol{I})\\ &= \dot{\alpha}\,\alpha\boldsymbol{I}.\end{aligned} \tag{4.105}$$

The rate of deformation can now be calculated from Equation (4.22) as

$$\boldsymbol{d} = \boldsymbol{F}^{-\mathrm{T}}\dot{\boldsymbol{E}}\boldsymbol{F}^{-1} = \frac{\dot{\alpha}}{\alpha}\boldsymbol{I}. \tag{4.106}$$

EXAMPLE 4.13: Textbook Exercise 4.10

Show that at the initial configuration, $\boldsymbol{F} = \boldsymbol{I}$, the linearization of $\hat{\boldsymbol{C}}$ in the direction of a displacement $\boldsymbol{u}$ is

$$D\hat{\boldsymbol{C}}[\boldsymbol{u}] = 2\boldsymbol{\varepsilon}' = 2\big[\boldsymbol{\varepsilon} - \tfrac{1}{3}(\mathrm{tr}\boldsymbol{\varepsilon})\boldsymbol{I}\big].$$

Solution

Recall from Equations (4.13) and (4.2) that

$$\hat{\boldsymbol{C}} = III_{\boldsymbol{C}}^{-\frac{1}{3}}\boldsymbol{C}, \tag{4.107}$$

where $\boldsymbol{C}$ is the right Cauchy–Green tensor and $III_{\boldsymbol{C}}$ is the third invariant (i.e., the determinant) of $\boldsymbol{C}$. Consequently,

$$D\hat{\boldsymbol{C}}[\boldsymbol{u}] = -\frac{1}{3}III_{\boldsymbol{C}}^{-\frac{4}{3}} DIII_{\boldsymbol{C}}[\boldsymbol{u}]\boldsymbol{C} + III_{\boldsymbol{C}}^{-\frac{1}{3}} D\boldsymbol{C}[\boldsymbol{u}]. \tag{4.108}$$

Noting that $III_{\boldsymbol{C}} = \det \boldsymbol{C} = J^2$ and from Equation (4.18) that $DJ[\boldsymbol{u}] = J\,\mathrm{tr}\boldsymbol{\varepsilon}$, enables the directional derivative $DIII_{\boldsymbol{C}}[\boldsymbol{u}]$ to be determined as

$$\begin{aligned} DIII_{\boldsymbol{C}}[\boldsymbol{u}] &= DJ^2[\boldsymbol{u}] \\ &= 2J\,DJ[\boldsymbol{u}] \\ &= 2J^2\,\mathrm{tr}\boldsymbol{\varepsilon} \\ &= 2III_{\boldsymbol{C}}(\mathrm{tr}\boldsymbol{\varepsilon}). \end{aligned} \tag{4.109}$$

Turning attention to the second directional derivative in Equation (4.108), observe that textbook Equation (4.74a) gives $D\boldsymbol{C}[\boldsymbol{u}] = 2\boldsymbol{F}^{\mathrm{T}}\boldsymbol{\varepsilon}\boldsymbol{F}$, which is proved using Equations (4.15) and (4.16) as follows:

$$\begin{aligned} D\boldsymbol{C}[\boldsymbol{u}] &= D(\boldsymbol{F}^{\mathrm{T}}\boldsymbol{F})[\boldsymbol{u}] \\ &= D\boldsymbol{F}^{\mathrm{T}}[\boldsymbol{u}]\boldsymbol{F} + \boldsymbol{F}^{\mathrm{T}}D\boldsymbol{F}[\boldsymbol{u}] \\ &= \boldsymbol{F}^{\mathrm{T}}(\boldsymbol{\nabla}\boldsymbol{u})\boldsymbol{F} + \boldsymbol{F}^{\mathrm{T}}(\boldsymbol{\nabla}\boldsymbol{u})^{\mathrm{T}}\boldsymbol{F} \\ &= \boldsymbol{F}^{\mathrm{T}}\left(\boldsymbol{\nabla}\boldsymbol{u} + (\boldsymbol{\nabla}\boldsymbol{u})^{\mathrm{T}}\right)\boldsymbol{F} \\ &= 2\boldsymbol{F}^{\mathrm{T}}\boldsymbol{\varepsilon}\boldsymbol{F}. \end{aligned} \tag{4.110}$$

Consequently Equation (4.108) can be rewritten as

$$\begin{aligned} D\hat{\boldsymbol{C}}[\boldsymbol{u}] &= -\frac{1}{3}III_{\boldsymbol{C}}^{-\frac{4}{3}}2III_{\boldsymbol{C}}(\mathrm{tr}\boldsymbol{\varepsilon})\boldsymbol{C} + III_{\boldsymbol{C}}^{-\frac{1}{3}}2\boldsymbol{F}^{\mathrm{T}}\boldsymbol{\varepsilon}\boldsymbol{F} \\ &= -\frac{2}{3}III_{\boldsymbol{C}}^{-\frac{1}{3}}(\mathrm{tr}\boldsymbol{\varepsilon})\boldsymbol{C} + 2III_{\boldsymbol{C}}^{-\frac{1}{3}}\boldsymbol{F}^{\mathrm{T}}\boldsymbol{\varepsilon}\boldsymbol{F}. \end{aligned} \tag{4.111}$$

At the initial configuration $\boldsymbol{F} = \boldsymbol{I}$, $\boldsymbol{C} = \boldsymbol{I}$ and $III_{\boldsymbol{C}} = 1$; hence

$$D\hat{\boldsymbol{C}}[\boldsymbol{u}] = -\frac{2}{3}\mathrm{tr}\boldsymbol{\varepsilon}\boldsymbol{I} + 2\boldsymbol{\varepsilon} = 2\left[\boldsymbol{\varepsilon} - \frac{1}{3}(\mathrm{tr}\boldsymbol{\varepsilon})\boldsymbol{I}\right]. \tag{4.112}$$

CHAPTER FIVE

STRESS AND EQUILIBRIUM

The examples presented in this chapter largely focus on the consequences of describing the virtual work expression of equilibrium in either a spatial or material configuration. An immediate result is the emergence of various alternative stress measures and the concept of work conjugacy. Examples involving stress rates and objective stress rates are also considered.

Equation summary

Vector projection interpretation of the tensor product [2.28]

$$(\boldsymbol{u} \otimes \boldsymbol{v})\boldsymbol{w} = (\boldsymbol{w} \cdot \boldsymbol{v})\boldsymbol{u}. \tag{5.1}$$

Double product $\boldsymbol{A} : \boldsymbol{B}$ in terms of the trace [2.51]

$$\begin{aligned} \boldsymbol{A} : \boldsymbol{B} &= \text{tr}(\boldsymbol{A}^T \boldsymbol{B}) = \text{tr}(\boldsymbol{B}\boldsymbol{A}^T) = \text{tr}(\boldsymbol{B}^T \boldsymbol{A}) \\ &= \text{tr}(\boldsymbol{A}\boldsymbol{B}^T) = \sum_{i,j=1}^{3} A_{ij} B_{ij}. \end{aligned} \tag{5.2}$$

Elemental area relation (Nanson's formula) [4.68]

$$d\boldsymbol{a} = J\boldsymbol{F}^{-T} d\boldsymbol{A}. \tag{5.3}$$

Virtual internal work as a function of the first Piola–Kirchhoff stress $\boldsymbol{P}$ [5.33]

$$\delta W_{\text{int}} = \int_V \boldsymbol{P} : \delta\dot{\boldsymbol{F}}\, dV \; ; \; \boldsymbol{P} = J\boldsymbol{\sigma}\boldsymbol{F}^{-T}. \tag{5.4}$$

Material differential equilibrium equation [5.36]

$$\nabla_0 \boldsymbol{P} + \boldsymbol{f}_0 = \boldsymbol{0}\,; \quad \nabla_0 \boldsymbol{P} = \frac{\partial \boldsymbol{P}}{\partial \boldsymbol{X}}. \tag{5.5}$$

Elemental force vector $d\boldsymbol{p}(\boldsymbol{\sigma})$ [5.38]

$$d\boldsymbol{p} = \boldsymbol{t}da = \boldsymbol{\sigma}d\boldsymbol{a}. \tag{5.6}$$

Elemental force vector $d\boldsymbol{p}(\boldsymbol{P})$ [5.39]

$$d\boldsymbol{p} = J\boldsymbol{\sigma}\boldsymbol{F}^{-T}d\boldsymbol{A} = \boldsymbol{P}d\boldsymbol{A}. \tag{5.7}$$

Cauchy stress $\boldsymbol{\sigma}$ in terms of first Piola–Kirchhoff stress $\boldsymbol{P}$ [5.45a]

$$\boldsymbol{\sigma} = J^{-1}\boldsymbol{P}\boldsymbol{F}^T. \tag{5.8}$$

Kirchhoff stress tensor $\boldsymbol{\tau} = J\boldsymbol{\sigma}$ as the push forward of second Piola–Kirchhoff stress $\boldsymbol{S}$ [5.46b]

$$\boldsymbol{\tau} = \boldsymbol{F}\boldsymbol{S}\boldsymbol{F}^T. \tag{5.9}$$

Time derivative of inverse of the deformation gradient $\boldsymbol{F}$ [5.55]

$$\frac{d}{dt}\boldsymbol{F}^{-1} = -\boldsymbol{F}^{-1}\boldsymbol{l}. \tag{5.10}$$

Deviatoric and pressure components of the Cauchy stress $\boldsymbol{\sigma}$ [5.49]

$$\boldsymbol{\sigma} = \boldsymbol{\sigma}' + p\boldsymbol{I}; \quad p = \frac{1}{3}\mathrm{tr}\boldsymbol{\sigma} = \frac{1}{3}\boldsymbol{\sigma} : \boldsymbol{I}. \tag{5.11}$$

Deviatoric and pressure components of the first Piola–Kirchhoff stress $\boldsymbol{P}$ [5.50a]

$$\boldsymbol{P} = \boldsymbol{P}' + pJ\boldsymbol{F}^{-T}; \quad \boldsymbol{P}' = J\boldsymbol{\sigma}'\boldsymbol{F}^{-T}. \tag{5.12}$$

Convective stress rate [5.60]

$$\begin{aligned}\boldsymbol{\sigma}^{\diamond} &= \boldsymbol{F}^{-T}\left[\frac{d}{dt}(\boldsymbol{F}^T\boldsymbol{\sigma}\boldsymbol{F})\right]\boldsymbol{F}^{-1} \\ &= \dot{\boldsymbol{\sigma}} + \boldsymbol{l}^T\boldsymbol{\sigma} + \boldsymbol{\sigma}\boldsymbol{l}.\end{aligned} \tag{5.13}$$

Jaumann stress rate, see [5.62]

$$\boldsymbol{\sigma}^{\nabla} = \dot{\boldsymbol{\sigma}} + \boldsymbol{\sigma}\boldsymbol{w} - \boldsymbol{w}\boldsymbol{\sigma}. \tag{5.14}$$

EXAMPLE 5.1

A three-dimensional finite deformation of a continuum from an initial configuration $\boldsymbol{X} = (X_1, X_2, X_3)^{\mathrm{T}}$ to a final configuration $\boldsymbol{x} = (x_1, x_2, x_3)^{\mathrm{T}}$

is given as

$$x_1(\boldsymbol{X}) = 5 - 3X_1 - X_2 \tag{5.15a}$$

$$x_2(\boldsymbol{X}) = 2 + \frac{5}{4}X_1 - 2X_2 \tag{5.15b}$$

$$x_3(\boldsymbol{X}) = X_3. \tag{5.15c}$$

In addition, the first Piola–Kirchhoff stress tensor is given by

$$\boldsymbol{P} = \begin{bmatrix} X_1 & X_1 & 0 \\ \alpha & X_2 & 0 \\ 0 & 0 & X_3 \end{bmatrix}. \tag{5.16}$$

(a) Determine if the deformation is isochoric.
(b) Determine the value of α so that the stress tensor $\boldsymbol{P}$ satisfies rotational equilibrium.
(c) Determine the body force field $\boldsymbol{f}_0$ per unit undeformed volume such that the material differential equilibrium Equation (5.5) is satisfied.

Solution

(a) From Equation (5.15) the deformation gradient is found as

$$\boldsymbol{F} = \begin{bmatrix} \partial x_1/\partial X_1 & \partial x_1/\partial X_2 & \partial x_1/\partial X_3 \\ \partial x_2/\partial X_1 & \partial x_2/\partial X_2 & \partial x_2/\partial X_3 \\ \partial x_3/\partial X_1 & \partial x_3/\partial X_2 & \partial x_3/\partial X_3 \end{bmatrix} = \begin{bmatrix} -3 & -1 & 0 \\ \frac{5}{4} & -2 & 0 \\ 0 & 0 & 1 \end{bmatrix}. \tag{5.17}$$

The volume ratio $J = \det \boldsymbol{F}$ is 7.25 which means the deformation is non-isochoric.
(b) From Equation (5.8), $\boldsymbol{\sigma} = J^{-1}\boldsymbol{P}\boldsymbol{F}^{\mathrm{T}}$ which is symmetric since $\boldsymbol{\sigma} = \boldsymbol{\sigma}^{\mathrm{T}}$; hence

$$J\boldsymbol{\sigma} = \begin{bmatrix} X_1 & X_1 & 0 \\ \alpha & X_2 & 0 \\ 0 & 0 & X_3 \end{bmatrix} \begin{bmatrix} -3 & \frac{5}{4} & 0 \\ -1 & -2 & 0 \\ 0 & 0 & 1 \end{bmatrix}$$

$$= \begin{bmatrix} -4X_1 & -\frac{3}{4}X_1 & 0 \\ -(3\alpha + X_2) & (\frac{5}{4}\alpha - 2X_2) & 0 \\ 0 & 0 & X_3 \end{bmatrix}. \tag{5.18}$$

Symmetry requires that

$$-\frac{3}{4}X_1 = -(3\alpha + X_2)\ ;\ \alpha = \frac{X_1}{4} - \frac{X_2}{3}, \tag{5.19}$$

giving

$$\boldsymbol{\sigma} = \frac{4}{29}\begin{bmatrix} -4X_1 & -\frac{3}{4}X_1 & 0 \\ -\frac{3}{4}X_1 & (\frac{5}{16}X_1 - \frac{29}{12}X_2) & 0 \\ 0 & 0 & X_3 \end{bmatrix}. \tag{5.20}$$

(c) From Equation (5.5) the material body force vector is

$$\boldsymbol{f}_0 = -\boldsymbol{\nabla}_0\boldsymbol{P} = \sum_{i,J=1}^{3} \frac{\partial P_{iJ}}{\partial X_J}, \tag{5.21}$$

where

$$\boldsymbol{P} = \begin{bmatrix} X_1 & X_1 & 0 \\ (\frac{X_1}{4} - \frac{X_2}{3}) & X_2 & 0 \\ 0 & 0 & X_3 \end{bmatrix}, \tag{5.22}$$

hence

$$\boldsymbol{f}_0 = -\begin{bmatrix} 1 \\ \frac{5}{4} \\ 1 \end{bmatrix}. \tag{5.23}$$

EXAMPLE 5.2

The deformation of a body is described by

$$x_1 = -3X_2\ ;\ x_2 = \frac{3}{2}X_1\ ;\ x_3 = X_3, \tag{5.24}$$

and the Cauchy stress tensor at a certain point in the spatial configuration is

$$\boldsymbol{\sigma} = \begin{bmatrix} 10 & 2 & 0 \\ 2 & 30 & 0 \\ 0 & 0 & 10 \end{bmatrix}. \tag{5.25}$$

Determine the Cauchy traction vector $\boldsymbol{t} = \boldsymbol{\sigma}\boldsymbol{n}$ and the first Piola–Kirchhoff traction vector $\boldsymbol{t}^{PK} = \boldsymbol{P}\boldsymbol{N}$ acting on a plane characterized by the spatial outward normal $\boldsymbol{n} = (0,1,0)^{\mathrm{T}}$.

Solution

From Equation (5.24) the basic kinematic deformation gradient quantities are easily found as

$$\boldsymbol{F} = \begin{bmatrix} 0 & -3 & 0 \\ \frac{3}{2} & 0 & 0 \\ 0 & 0 & 1 \end{bmatrix} ; \ \boldsymbol{F}^{-1} = \begin{bmatrix} 0 & \frac{2}{3} & 0 \\ -\frac{1}{3} & 0 & 0 \\ 0 & 0 & 1 \end{bmatrix}, \tag{5.26}$$

$$\boldsymbol{F}^{\mathrm{T}} = \begin{bmatrix} 0 & \frac{3}{2} & 0 \\ -3 & 0 & 0 \\ 0 & 0 & 1 \end{bmatrix} ; \ \boldsymbol{F}^{-\mathrm{T}} = \begin{bmatrix} 0 & -\frac{1}{3} & 0 \\ \frac{2}{3} & 0 & 0 \\ 0 & 0 & 1 \end{bmatrix}. \tag{5.27}$$

From Equation (5.26) the elemental volume ratio $J = \det \boldsymbol{F} = 9/2$. The Cauchy stress tensor is trivially found as

$$\boldsymbol{t} = \boldsymbol{\sigma}\boldsymbol{n} = \begin{bmatrix} 10 & 2 & 0 \\ 2 & 30 & 0 \\ 0 & 0 & 10 \end{bmatrix} \begin{bmatrix} 0 \\ 1 \\ 0 \end{bmatrix} = \begin{bmatrix} 2 \\ 30 \\ 0 \end{bmatrix}. \tag{5.28}$$

From Nanson's formula given by Equation (5.3) the material unit normal $\boldsymbol{N}$ is the pull back of the spatial unit normal $\boldsymbol{n}$ as

$$\boldsymbol{N} = \left(\frac{J^{-1}da}{dA}\right)\boldsymbol{F}^{\mathrm{T}}\boldsymbol{n}, \tag{5.29}$$

where da and dA are now scalars. Insofar as $\boldsymbol{N}$ is a unit vector

$$\boldsymbol{N} = \frac{\boldsymbol{F}^{\mathrm{T}}\boldsymbol{n}}{\|\boldsymbol{F}^{\mathrm{T}}\boldsymbol{n}\|}. \tag{5.30}$$

For the spatial unit normal $\boldsymbol{n} = (0,1,0)^{\mathrm{T}}$ it is easy to find $\boldsymbol{N} = (1,0,0)^{\mathrm{T}}$. Using Equation (5.8) the first Piola–Kirchhoff stress is found as

$$\begin{aligned}\boldsymbol{P} &= J\boldsymbol{\sigma}\boldsymbol{F}^{-\mathrm{T}}\\ &= \frac{9}{2}\begin{bmatrix}10 & 2 & 0\\ 2 & 30 & 0\\ 0 & 0 & 10\end{bmatrix}\begin{bmatrix}0 & -\frac{1}{3} & 0\\ \frac{2}{3} & 0 & 0\\ 0 & 0 & 1\end{bmatrix}\\ &= \frac{9}{2}\begin{bmatrix}\frac{4}{3} & -\frac{10}{3} & 0\\ 20 & -\frac{2}{3} & 0\\ 0 & 0 & 10\end{bmatrix}.\end{aligned} \tag{5.31}$$

Consequently for $\boldsymbol{N} = (1,0,0)^{\mathrm{T}}$

$$\boldsymbol{t}^{PK} = \boldsymbol{P}\boldsymbol{N} = \frac{9}{2}\begin{bmatrix}\frac{4}{3}\\ 20\\ 0\end{bmatrix} = \begin{bmatrix}6\\ 90\\ 0\end{bmatrix}. \tag{5.32}$$

It is worth exploring this example a little more since both $\boldsymbol{\sigma}$ and $\boldsymbol{P}$ should yield the same elemental force vector $d\boldsymbol{p}$ provided they are post multiplied by the spatial and material elemental areas $d\boldsymbol{a}$ and $d\boldsymbol{A}$, respectively. To this end Equation (4.86) is employed to calculate the area ratio using the material unit vector $\boldsymbol{N}$ associated with area dA and the inverse of the right Cauchy–Green deformation tensor $\boldsymbol{C}^{-1}$ as

$$\frac{da}{dA} = J\sqrt{\boldsymbol{N}\cdot\boldsymbol{C}^{-1}\boldsymbol{N}}\,; \quad \boldsymbol{C}^{-1} = \boldsymbol{F}^{-1}\boldsymbol{F}^{-\mathrm{T}}. \tag{5.33}$$

For $\boldsymbol{N} = (1,0,0)^{\mathrm{T}}$ and using Equations (5.26) and (5.27), the area ratio is calculated as

$$\frac{da}{dA} = 3. \tag{5.34}$$

Using Equations (5.6), (5.7), (5.28), and (5.32), the elemental force vector $d\boldsymbol{p}$ can be variously calculated in terms of the Cauchy stress $\boldsymbol{\sigma}$ and first

Piola–Kirchhoff stress $\boldsymbol{P}$ as

$$d\boldsymbol{p} = \boldsymbol{\sigma}\boldsymbol{n}da = \begin{bmatrix} 10 & 2 & 0 \\ 2 & 30 & 0 \\ 0 & 0 & 10 \end{bmatrix} \begin{bmatrix} 0 \\ 1 \\ 0 \end{bmatrix} da = \begin{bmatrix} 2 \\ 30 \\ 0 \end{bmatrix} da. \tag{5.35}$$

Or

$$d\boldsymbol{p} = \boldsymbol{P}\boldsymbol{N}dA = \frac{9}{2}\begin{bmatrix} \frac{4}{3} & -\frac{10}{3} & 0 \\ 20 & -\frac{2}{3} & 0 \\ 0 & 0 & 10 \end{bmatrix} \begin{bmatrix} 1 \\ 0 \\ 0 \end{bmatrix} dA = \begin{bmatrix} 6 \\ 90 \\ 0 \end{bmatrix} dA. \tag{5.36}$$

Observing that $dA = da/3$ shows that the same elemental force can be calculated using either a spatial or material description.

EXAMPLE 5.3: Textbook Exercise 5.1

A two-dimensional Cauchy stress tensor is given as

$$\boldsymbol{\sigma} = \boldsymbol{t} \otimes \boldsymbol{n}_1 + \alpha\, \boldsymbol{n}_1 \otimes \boldsymbol{n}_2,$$

where $\boldsymbol{t}$ is an arbitrary vector and $\boldsymbol{n}_1$ and $\boldsymbol{n}_2$ are orthogonal unit vectors.
(a) Describe graphically the state of stress.
(b) Determine the value of α. (Hint: $\boldsymbol{\sigma}$ must be symmetric.)

Solution

To reveal the two-dimensional nature of the stress tensor, investigate the traction vector existing on planes normal to the two orthogonal axes. This is achieved by multiplying $\boldsymbol{\sigma}$, in turn, by $\boldsymbol{n}_1$ and $\boldsymbol{n}_2$ and employing Equation (5.1):

$$\begin{aligned} \boldsymbol{\sigma}\boldsymbol{n}_1 &= (\boldsymbol{t} \otimes \boldsymbol{n}_1)\boldsymbol{n}_1 + \alpha(\boldsymbol{n}_1 \otimes \boldsymbol{n}_2)\boldsymbol{n}_1 \\ &= (\boldsymbol{n}_1 \cdot \boldsymbol{n}_1)\,\boldsymbol{t} + \alpha(\boldsymbol{n}_1 \cdot \boldsymbol{n}_2)\,\boldsymbol{n}_1 \\ &= \boldsymbol{t}. \end{aligned} \tag{5.37}$$

$$\begin{aligned} \boldsymbol{\sigma}\boldsymbol{n}_2 &= (\boldsymbol{t} \otimes \boldsymbol{n}_1)\boldsymbol{n}_2 + \alpha(\boldsymbol{n}_1 \otimes \boldsymbol{n}_2)\boldsymbol{n}_2 \\ &= (\boldsymbol{n}_2 \cdot \boldsymbol{n}_1)\,\boldsymbol{t} + \alpha(\boldsymbol{n}_2 \cdot \boldsymbol{n}_2)\boldsymbol{n}_1 \\ &= \alpha\,\boldsymbol{n}_1. \end{aligned} \tag{5.38}$$

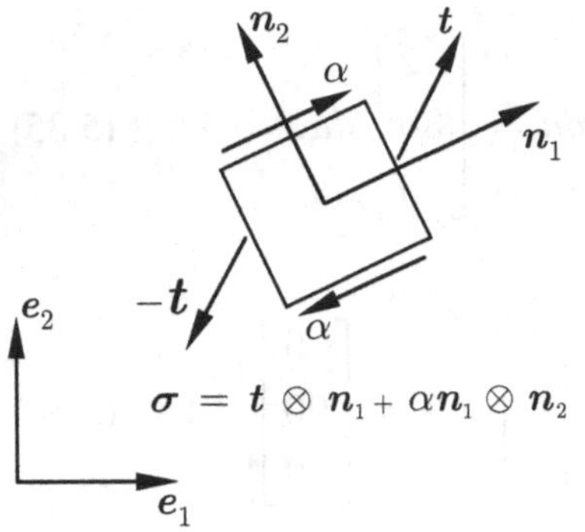

FIGURE 5.1 Example 5.3

Consequently the traction vector $\boldsymbol{t}$ operates on the surface normal to $\boldsymbol{n}_1$ while a shear force of magnitude α acts in the $\boldsymbol{n}_1$ direction in the surface normal to $\boldsymbol{n}_2$, see Figure 5.1. It is easy to show that since $\boldsymbol{\sigma}$ is symmetric then $\boldsymbol{n}_1 \cdot \boldsymbol{\sigma}\boldsymbol{n}_2 = \boldsymbol{n}_2 \cdot \boldsymbol{\sigma}\boldsymbol{n}_1$. Expanding these terms enables α to be found as follows:

$$\begin{aligned} \boldsymbol{n}_1 \cdot \boldsymbol{\sigma}\boldsymbol{n}_2 &= \boldsymbol{n}_1 \cdot (\boldsymbol{t} \otimes \boldsymbol{n}_1)\boldsymbol{n}_2 + \alpha\boldsymbol{n}_1 \cdot (\boldsymbol{n}_1 \otimes \boldsymbol{n}_2)\boldsymbol{n}_2 \\ &= \boldsymbol{n}_1 \cdot (\boldsymbol{n}_2 \cdot \boldsymbol{n}_1)\boldsymbol{t} + \alpha\boldsymbol{n}_1 \cdot (\boldsymbol{n}_2 \cdot \boldsymbol{n}_2)\boldsymbol{n}_1 \\ &= \alpha. \end{aligned} \tag{5.39}$$

$$\begin{aligned} \boldsymbol{n}_2 \cdot \boldsymbol{\sigma}\boldsymbol{n}_1 &= \boldsymbol{n}_2 \cdot (\boldsymbol{t} \otimes \boldsymbol{n}_1)\boldsymbol{n}_1 + \alpha\boldsymbol{n}_2 \cdot (\boldsymbol{n}_1 \otimes \boldsymbol{n}_2)\boldsymbol{n}_1 \\ &= \boldsymbol{n}_2 \cdot (\boldsymbol{n}_1 \cdot \boldsymbol{n}_1)\boldsymbol{t} + \alpha\boldsymbol{n}_2 \cdot (\boldsymbol{n}_1 \cdot \boldsymbol{n}_2)\boldsymbol{n}_1 \\ &= \boldsymbol{n}_2 \cdot \boldsymbol{t}. \end{aligned} \tag{5.40}$$

This gives $\alpha = \boldsymbol{n}_2 \cdot \boldsymbol{t}$ which is also the magnitude of the component of $\boldsymbol{t}$ in the direction $\boldsymbol{n}_2$ required for symmetry of the Cauchy stress tensor.

EXAMPLE 5.4: Textbook Exercise 5.2

Using Equation (5.10) and a process similar to that employed in textbook Example 5.5, page 150, show that, with respect to the initial volume, the stress tensor $\boldsymbol{\Pi}$ is work conjugate to the tensor $\dot{\boldsymbol{H}}$, where $\boldsymbol{H} = -\boldsymbol{F}^{-\mathrm{T}}$ and $\boldsymbol{\Pi} = \boldsymbol{P}\boldsymbol{C} = J\boldsymbol{\sigma}\boldsymbol{F}$.

Solution

Observe from Equation (5.4) that the first Piola–Kirchhoff stress tensor $\boldsymbol{P}$ is work conjugate to the time derivative of the deformation gradient $\boldsymbol{F}$. Consequently the work term $\boldsymbol{P} : \dot{\boldsymbol{F}}$ will provide a vehicle for addressing the question. The term $\dot{\boldsymbol{F}}$ can be found by taking the time derivative of the identity $\boldsymbol{F}\boldsymbol{F}^{-1} = \boldsymbol{I}$ as:

$$\frac{d}{dt}(\boldsymbol{F}\boldsymbol{F}^{-1}) = \dot{\boldsymbol{F}}\boldsymbol{F}^{-1} + \boldsymbol{F}\dot{\boldsymbol{F}}^{-1} = \boldsymbol{0}, \tag{5.41}$$

hence

$$\dot{\boldsymbol{F}} = -\boldsymbol{F}\dot{\boldsymbol{F}}^{-1}\boldsymbol{F}. \tag{5.42}$$

The first Piola–Kirchhoff stress tensor is $\boldsymbol{P} = J\boldsymbol{\sigma}\boldsymbol{F}^{-\mathrm{T}}$ which together with the above equation and the properties of the trace given by Equation (5.2) gives

$$\begin{aligned}
\boldsymbol{P} : \dot{\boldsymbol{F}} &= -\left(J\boldsymbol{\sigma}\boldsymbol{F}^{-\mathrm{T}}\right) : \left(\boldsymbol{F}\dot{\boldsymbol{F}}^{-1}\boldsymbol{F}\right) \\
&= -J\mathrm{tr}\left(\boldsymbol{\sigma}\boldsymbol{F}^{-\mathrm{T}}(\boldsymbol{F}\dot{\boldsymbol{F}}^{-1}\boldsymbol{F})^{\mathrm{T}}\right) \\
&= -J\mathrm{tr}\left(\boldsymbol{\sigma}\boldsymbol{F}^{-\mathrm{T}}\boldsymbol{F}^{\mathrm{T}}\dot{\boldsymbol{F}}^{-\mathrm{T}}\boldsymbol{F}^{\mathrm{T}}\right) \\
&= -J\mathrm{tr}\left(\boldsymbol{\sigma}(\dot{\boldsymbol{F}}^{-\mathrm{T}}\boldsymbol{F}^{\mathrm{T}})\right) \\
&= -J\mathrm{tr}\left((\dot{\boldsymbol{F}}^{-\mathrm{T}}(\boldsymbol{F}^{\mathrm{T}}\boldsymbol{\sigma})\right) \\
&= -J\mathrm{tr}\left(\dot{\boldsymbol{F}}^{-\mathrm{T}}(\boldsymbol{\sigma}\boldsymbol{F})^{\mathrm{T}}\right) \\
&= -J\boldsymbol{\sigma}\boldsymbol{F} : \dot{\boldsymbol{F}}^{-\mathrm{T}} \\
&= \boldsymbol{\Pi} : \dot{\boldsymbol{H}},
\end{aligned} \tag{5.43}$$

where use has been made of the symmetry of $\boldsymbol{\sigma}$. Equation (5.43) shows that $\boldsymbol{\Pi}$ is work conjugate to $\dot{\boldsymbol{H}}$. Using Equation (5.8) it is easy to show that $\boldsymbol{P}\boldsymbol{C} = J\boldsymbol{\sigma}\boldsymbol{F}$.

EXAMPLE 5.5: Textbook Exercise 5.3

Using the time derivative of the equality $\boldsymbol{C}\boldsymbol{C}^{-1} = \boldsymbol{I}$, show that the tensor $\boldsymbol{\Sigma} = \boldsymbol{C}\boldsymbol{S}\,\boldsymbol{C} = J\boldsymbol{F}^{\mathrm{T}}\boldsymbol{\sigma}\boldsymbol{F}$ is work conjugate to $\frac{1}{2}\dot{\boldsymbol{B}}$, where $\boldsymbol{B} = -\boldsymbol{C}^{-1}$ with respect to the initial volume. Find relationships between $\boldsymbol{T}$ (Biot stress tensor), $\boldsymbol{\Sigma}$, and $\boldsymbol{\Pi}$.

Solution

The time derivative of $\boldsymbol{C}\boldsymbol{C}^{-1} = \boldsymbol{I}$ is simply found to be

$$\frac{d}{dt}(\boldsymbol{C}\boldsymbol{C}^{-1}) = \dot{\boldsymbol{C}}\boldsymbol{C}^{-1} + \boldsymbol{C}\dot{\boldsymbol{C}}^{-1} = \boldsymbol{0}, \tag{5.44}$$

hence

$$\dot{\boldsymbol{C}} = -\boldsymbol{C}\dot{\boldsymbol{C}}^{-1}\boldsymbol{C}. \tag{5.45}$$

A starting point can be found by observing that the second Piola–Kirchhoff stress tensor $\boldsymbol{S}$ is work conjugate to the time derivative of the Green's strain $\boldsymbol{E}$ with respect to the initial volume and that $\dot{\boldsymbol{E}} = \frac{1}{2}\dot{\boldsymbol{C}}$ where $\boldsymbol{C}$ is the right Cauchy–Green tensor. Hence,

$$\boldsymbol{S} : \dot{\boldsymbol{E}} = \frac{1}{2}\boldsymbol{S} : \dot{\boldsymbol{C}}. \tag{5.46}$$

Substituting from Equation (5.45) for $\dot{\boldsymbol{C}}$ and using the properties of the trace given by Equation (5.2) together with symmetries of $\boldsymbol{S}$ and $\boldsymbol{C}$ enables the work conjugate relationship between $\boldsymbol{\Sigma}$ and $\frac{1}{2}\dot{\boldsymbol{B}}$ to be found:

$$\begin{aligned}
\boldsymbol{S} : \dot{\boldsymbol{E}} &= -\frac{1}{2}\boldsymbol{S} : (\boldsymbol{C}\dot{\boldsymbol{C}}^{-1}\boldsymbol{C}) \\
&= -\frac{1}{2}\mathrm{tr}(\boldsymbol{S}^{\mathrm{T}}\boldsymbol{C}\dot{\boldsymbol{C}}^{-1}\boldsymbol{C}) \\
&= -\frac{1}{2}\mathrm{tr}((\boldsymbol{S}\boldsymbol{C}\dot{\boldsymbol{C}}^{-1})\,\boldsymbol{C}^{\mathrm{T}}) \\
&= (\boldsymbol{C}\boldsymbol{S}\,\boldsymbol{C}) : \left(-\frac{1}{2}\dot{\boldsymbol{C}}^{-1}\right) \\
&= \boldsymbol{\Pi} : \left(\frac{1}{2}\dot{\boldsymbol{B}}\right).
\end{aligned} \tag{5.47}$$

From textbook Example 5.5 (page 150) the Biot stress tensor is

$$\boldsymbol{T} = \frac{1}{2}(\boldsymbol{S}\boldsymbol{U} + \boldsymbol{U}\boldsymbol{S}). \tag{5.48}$$

Noting the following relationships

$$\boldsymbol{F} = \boldsymbol{R}\boldsymbol{U}\ ; \quad \boldsymbol{U} = \boldsymbol{R}^{\mathrm{T}}\boldsymbol{F}, \tag{5.49}$$

$$\boldsymbol{P} = \boldsymbol{F}\boldsymbol{S}\ ; \quad \boldsymbol{S} = \boldsymbol{F}^{-1}\boldsymbol{P}, \tag{5.50}$$

and that $\boldsymbol{S}$ and $\boldsymbol{U}$ are symmetric enables the Biot stress to be expressed as

$$\begin{aligned}\boldsymbol{T} &= \frac{1}{2}((\boldsymbol{P}^{\mathrm{T}}\boldsymbol{F}^{-\mathrm{T}})(\boldsymbol{R}^{\mathrm{T}}\boldsymbol{F}) + (\boldsymbol{R}^{\mathrm{T}}\boldsymbol{F})(\boldsymbol{F}^{-1}\boldsymbol{P}))\\ &= \frac{1}{2}(\boldsymbol{P}^{\mathrm{T}}\boldsymbol{F}^{-\mathrm{T}}\boldsymbol{F}^{\mathrm{T}}\boldsymbol{R} + \boldsymbol{R}^{\mathrm{T}}\boldsymbol{F}\boldsymbol{F}^{-1}\boldsymbol{P})\\ &= \frac{1}{2}(\boldsymbol{P}^{\mathrm{T}}\boldsymbol{R} + \boldsymbol{R}^{\mathrm{T}}\boldsymbol{P}).\end{aligned} \tag{5.51}$$

From the previous example, $\boldsymbol{\Pi} = \boldsymbol{P}\boldsymbol{C}$, hence

$$\boldsymbol{T} = \frac{1}{2}(\boldsymbol{C}^{-1}\boldsymbol{\Pi}^{\mathrm{T}}\boldsymbol{R} + \boldsymbol{R}^{\mathrm{T}}\boldsymbol{\Pi}\boldsymbol{C}^{-1}). \tag{5.52}$$

We can now find $\boldsymbol{\Pi}$ in terms of $\boldsymbol{\Sigma}$ using Equation (5.50) as follows:

$$\begin{aligned}\boldsymbol{\Sigma} &= \boldsymbol{C}\boldsymbol{S}\boldsymbol{C}\\ &= \boldsymbol{C}\boldsymbol{F}^{-1}\boldsymbol{P}\boldsymbol{C}\\ &= \boldsymbol{F}^{\mathrm{T}}\boldsymbol{F}\boldsymbol{F}^{-1}\boldsymbol{\Pi}\boldsymbol{C}^{-1}\boldsymbol{C}\\ &= \boldsymbol{F}^{\mathrm{T}}\boldsymbol{\Pi}.\end{aligned} \tag{5.53}$$

Recalling that $\boldsymbol{\Sigma}$ is symmetric gives

$$\boldsymbol{\Pi} = \boldsymbol{F}^{-\mathrm{T}}\boldsymbol{\Sigma}\ ; \quad \boldsymbol{\Pi}^{\mathrm{T}} = \boldsymbol{\Sigma}\boldsymbol{F}^{-1}. \tag{5.54}$$

Substituting into Equation (5.52) gives

$$\boldsymbol{T} = \frac{1}{2}(\boldsymbol{C}^{-1}\boldsymbol{\Sigma}\boldsymbol{F}^{-1}\boldsymbol{R} + \boldsymbol{R}^{\mathrm{T}}\boldsymbol{F}^{-\mathrm{T}}\boldsymbol{\Sigma}\boldsymbol{C}^{-1}). \tag{5.55}$$

Again noting that $\boldsymbol{U}$ is symmetric and that $\boldsymbol{F} = \boldsymbol{R}\boldsymbol{U}$ gives

$$\boldsymbol{T} = \frac{1}{2}(\boldsymbol{C}^{-1}\boldsymbol{\Sigma}\boldsymbol{U}^{-1} + \boldsymbol{U}^{-1}\boldsymbol{\Sigma}\boldsymbol{C}^{-1}). \tag{5.56}$$

EXAMPLE 5.6: Textbook Exercise 5.4

Prove $\boldsymbol{P}' : \boldsymbol{F} = 0$ using a procedure similar to textbook Example 5.6, page 152, where $\boldsymbol{P}'$ is the deviatoric component of the first Piola–Kirchhoff stress tensor and $\boldsymbol{F}$ if the deformation gradient tensor.

Solution

From Equations (5.11) and (5.12) the deviatoric first Piola–Kirchhoff stress tensor can be written as

$$\begin{aligned}\boldsymbol{P}' &= J\boldsymbol{\sigma}'\boldsymbol{F}^{-\mathrm{T}} \\ &= J(\boldsymbol{\sigma} - p\boldsymbol{I})\boldsymbol{F}^{-\mathrm{T}} \\ &= J(\boldsymbol{\sigma} - \tfrac{1}{3}\mathrm{tr}\boldsymbol{\sigma}\boldsymbol{I})\boldsymbol{F}^{-\mathrm{T}}.\end{aligned} \tag{5.57}$$

We can therefore expand $\boldsymbol{P}' : \boldsymbol{F} = 0$ as

$$\begin{aligned}\boldsymbol{P}' : \boldsymbol{F} &= J(\boldsymbol{\sigma}\boldsymbol{F}^{-\mathrm{T}}) : \boldsymbol{F} - \tfrac{1}{3}J\,\mathrm{tr}\boldsymbol{\sigma}(\boldsymbol{I}\boldsymbol{F}^{-\mathrm{T}}) : \boldsymbol{F} \\ &= J\mathrm{tr}\big((\boldsymbol{F}^{-1}\boldsymbol{\sigma})\boldsymbol{F}\big) - \tfrac{1}{3}J\,\mathrm{tr}\boldsymbol{\sigma}\mathrm{tr}(\boldsymbol{F}^{-1}\boldsymbol{I}\boldsymbol{F}) \\ &= J\mathrm{tr}(\boldsymbol{F}\boldsymbol{F}^{-1}\boldsymbol{\sigma}) - \tfrac{1}{3}J\,\mathrm{tr}\boldsymbol{\sigma}\mathrm{tr}(\boldsymbol{I}) \\ &= J\mathrm{tr}(\boldsymbol{\sigma}) - \tfrac{1}{3}J\,\mathrm{tr}\boldsymbol{\sigma}\,(3) \\ &= 0.\end{aligned} \tag{5.58}$$

EXAMPLE 5.7: Textbook Exercise 5.5

Prove directly that the Jaumann stress (rate) tensor, $\boldsymbol{\sigma}^{\triangledown}$, is an objective tensor, using a procedure similar to textbook Example 5.7, page 154.

Solution

Recall from Equation (5.14) that the Jaumann stress rate tensor is given by $\boldsymbol{\sigma}^{\triangledown} = \dot{\boldsymbol{\sigma}} + \boldsymbol{\sigma}\boldsymbol{w} - \boldsymbol{w}\boldsymbol{\sigma}$ and that for $\boldsymbol{\sigma}^{\triangledown}$ to be objective under the action of a superimposed rigid body motion $\boldsymbol{Q}$, then $\tilde{\boldsymbol{\sigma}}^{\triangledown} = \boldsymbol{Q}\boldsymbol{\sigma}^{\triangledown}\boldsymbol{Q}^{\mathrm{T}}$.
The rotated Jaumann stress rate tensor is

$$\tilde{\boldsymbol{\sigma}}^{\triangledown} = \dot{\tilde{\boldsymbol{\sigma}}} + \tilde{\boldsymbol{\sigma}}\tilde{\boldsymbol{w}} - \tilde{\boldsymbol{w}}\tilde{\boldsymbol{\sigma}}, \tag{5.59}$$

where, from textbook Equations (5.11), (5.53), and (4.108)

$$\tilde{\boldsymbol{\sigma}} = \boldsymbol{Q}\boldsymbol{\sigma}\boldsymbol{Q}^{\mathrm{T}} \tag{5.60a}$$

$$\dot{\tilde{\boldsymbol{\sigma}}} = \boldsymbol{Q}\dot{\boldsymbol{\sigma}}\boldsymbol{Q}^{\mathrm{T}} + \dot{\boldsymbol{Q}}\boldsymbol{\sigma}\boldsymbol{Q}^{\mathrm{T}} + \boldsymbol{Q}\boldsymbol{\sigma}\dot{\boldsymbol{Q}}^{\mathrm{T}} \tag{5.60b}$$

$$\tilde{\boldsymbol{w}} = \tfrac{1}{2}(\tilde{\boldsymbol{l}} - \tilde{\boldsymbol{l}}^{\mathrm{T}}), \tag{5.60c}$$

and where textbook Equations (4.108) and (4.137) give

$$\tilde{\boldsymbol{l}} = \boldsymbol{Q}\boldsymbol{l}\boldsymbol{Q}^{\mathrm{T}} + \dot{\boldsymbol{Q}}\boldsymbol{Q}^{\mathrm{T}} \tag{5.61a}$$

$$\tilde{\boldsymbol{l}}^{\mathrm{T}} = \boldsymbol{Q}\boldsymbol{l}^{\mathrm{T}}\boldsymbol{Q}^{\mathrm{T}} + \boldsymbol{Q}\dot{\boldsymbol{Q}}^{\mathrm{T}} \tag{5.61b}$$

$$\tilde{\boldsymbol{w}} = \boldsymbol{Q}\boldsymbol{w}\boldsymbol{Q}^{\mathrm{T}} + \tfrac{1}{2}(\dot{\boldsymbol{Q}}\boldsymbol{Q}^{\mathrm{T}} - \boldsymbol{Q}\dot{\boldsymbol{Q}}^{\mathrm{T}}). \tag{5.61c}$$

Substitute Equations (5.60a–5.60c) and (5.61a–5.61c) into Equation (5.59) and note that $\boldsymbol{Q}\boldsymbol{Q}^{\mathrm{T}} = \boldsymbol{I}$ to give

$$\begin{aligned}
\tilde{\boldsymbol{\sigma}}^{\nabla} &= \boldsymbol{Q}\dot{\boldsymbol{\sigma}}\boldsymbol{Q}^{\mathrm{T}} + \dot{\boldsymbol{Q}}\boldsymbol{\sigma}\boldsymbol{Q}^{\mathrm{T}} + \boldsymbol{Q}\boldsymbol{\sigma}\dot{\boldsymbol{Q}}^{\mathrm{T}} \\
&\quad + \boldsymbol{Q}\boldsymbol{\sigma}\boldsymbol{Q}^{\mathrm{T}}\left(\boldsymbol{Q}\boldsymbol{w}\boldsymbol{Q}^{\mathrm{T}} + \tfrac{1}{2}(\dot{\boldsymbol{Q}}\boldsymbol{Q}^{\mathrm{T}} - \boldsymbol{Q}\dot{\boldsymbol{Q}}^{\mathrm{T}})\right) \\
&\quad - \left(\boldsymbol{Q}\boldsymbol{w}\boldsymbol{Q}^{\mathrm{T}} + \tfrac{1}{2}(\dot{\boldsymbol{Q}}\boldsymbol{Q}^{\mathrm{T}} - \boldsymbol{Q}\dot{\boldsymbol{Q}}^{\mathrm{T}})\right)\boldsymbol{Q}\boldsymbol{\sigma}\boldsymbol{Q}^{\mathrm{T}} \\
&= \underline{\boldsymbol{Q}\dot{\boldsymbol{\sigma}}\boldsymbol{Q}^{\mathrm{T}}} + \dot{\boldsymbol{Q}}\boldsymbol{\sigma}\boldsymbol{Q}^{\mathrm{T}} + \boldsymbol{Q}\boldsymbol{\sigma}\dot{\boldsymbol{Q}}^{\mathrm{T}} \\
&\quad + \underline{\boldsymbol{Q}\boldsymbol{\sigma}\boldsymbol{w}\boldsymbol{Q}^{\mathrm{T}}} + \tfrac{1}{2}\boldsymbol{Q}\boldsymbol{\sigma}\boldsymbol{Q}^{\mathrm{T}}\dot{\boldsymbol{Q}}\boldsymbol{Q}^{\mathrm{T}} - \tfrac{1}{2}\boldsymbol{Q}\boldsymbol{\sigma}\dot{\boldsymbol{Q}}^{\mathrm{T}} \\
&\quad - \underline{\boldsymbol{Q}\boldsymbol{w}\boldsymbol{\sigma}\boldsymbol{Q}^{\mathrm{T}}} - \tfrac{1}{2}\dot{\boldsymbol{Q}}\boldsymbol{\sigma}\boldsymbol{Q}^{\mathrm{T}} + \tfrac{1}{2}\boldsymbol{Q}\dot{\boldsymbol{Q}}^{\mathrm{T}}\boldsymbol{Q}\boldsymbol{\sigma}\boldsymbol{Q}^{\mathrm{T}}.
\end{aligned} \tag{5.62}$$

Since $\boldsymbol{Q}\boldsymbol{Q}^{\mathrm{T}} = \boldsymbol{I}$,

$$\dot{\boldsymbol{Q}}\boldsymbol{Q}^{\mathrm{T}} + \boldsymbol{Q}\dot{\boldsymbol{Q}}^{\mathrm{T}} = 0. \tag{5.63a}$$

$$\dot{\boldsymbol{Q}}^{\mathrm{T}} = -\boldsymbol{Q}^{\mathrm{T}}\dot{\boldsymbol{Q}}\boldsymbol{Q}^{\mathrm{T}}. \tag{5.63b}$$

Observe that the underlined elements in Equation (5.62) constitute the objective measure of the Jaumann stress tensor: that is,

$$\boldsymbol{Q}\dot{\boldsymbol{\sigma}}\boldsymbol{Q}^{\mathrm{T}} + \boldsymbol{Q}\boldsymbol{\sigma}\boldsymbol{w}\boldsymbol{Q}^{\mathrm{T}} - \boldsymbol{Q}\boldsymbol{w}\boldsymbol{\sigma}\boldsymbol{Q}^{\mathrm{T}} = \boldsymbol{Q}\boldsymbol{\sigma}^{\nabla}\boldsymbol{Q}^{\mathrm{T}}. \tag{5.64}$$

Now substituting for $\dot{\boldsymbol{Q}}^{\mathrm{T}}$ in Equation (5.62) gives

$$
\begin{aligned}
\tilde{\boldsymbol{\sigma}}^{\nabla} &= \boldsymbol{Q}\boldsymbol{\sigma}^{\nabla}\boldsymbol{Q}^{\mathrm{T}} + \dot{\boldsymbol{Q}}\boldsymbol{\sigma}\boldsymbol{Q}^{\mathrm{T}} - \boldsymbol{Q}\boldsymbol{\sigma}\boldsymbol{Q}^{\mathrm{T}}\dot{\boldsymbol{Q}}\boldsymbol{Q}^{\mathrm{T}} \\
&\quad + \tfrac{1}{2}\boldsymbol{Q}\boldsymbol{\sigma}\boldsymbol{Q}^{\mathrm{T}}\dot{\boldsymbol{Q}}\boldsymbol{Q}^{\mathrm{T}} + \tfrac{1}{2}\boldsymbol{Q}\boldsymbol{\sigma}\boldsymbol{Q}^{\mathrm{T}}\dot{\boldsymbol{Q}}\boldsymbol{Q}^{\mathrm{T}} \\
&\quad - \tfrac{1}{2}\dot{\boldsymbol{Q}}\boldsymbol{\sigma}\boldsymbol{Q}^{\mathrm{T}} - \tfrac{1}{2}\boldsymbol{Q}\boldsymbol{Q}^{\mathrm{T}}\dot{\boldsymbol{Q}}\boldsymbol{Q}^{\mathrm{T}}\boldsymbol{Q}\boldsymbol{\sigma}\boldsymbol{Q}^{\mathrm{T}} \\
&= \boldsymbol{Q}\boldsymbol{\sigma}^{\nabla}\boldsymbol{Q}^{\mathrm{T}},
\end{aligned}
\tag{5.65}
$$

thus proving the objectivity of the Jaumann stress rate tensor.

EXAMPLE 5.8: Textbook Exercise 5.6

Prove that if $d\boldsymbol{x}_1$ and $d\boldsymbol{x}_2$ are two arbitrary elemental vectors moving with the body (see textbook Figure 4.2), then

$$\frac{d}{dt}(d\boldsymbol{x}_1 \cdot \boldsymbol{\sigma} d\boldsymbol{x}_2) = d\boldsymbol{x}_1 \cdot \boldsymbol{\sigma}^{\diamond} d\boldsymbol{x}_2,$$

where we recall from Equation (5.13) that $\boldsymbol{\sigma}^{\diamond}$ is the convective stress rate given by

$$\boldsymbol{\sigma}^{\diamond} = \dot{\boldsymbol{\sigma}} + \boldsymbol{l}^{\mathrm{T}}\boldsymbol{\sigma} + \boldsymbol{\sigma}\boldsymbol{l}. \tag{5.66}$$

Solution

When dealing with time derivatives of spatial vectors such as $d\boldsymbol{x}_1$ and $d\boldsymbol{x}_2$ it is convenient to re-express them in terms of the equivalent material vectors as $d\boldsymbol{x}_1 = \boldsymbol{F}d\boldsymbol{X}_1$ and $d\boldsymbol{x}_2 = \boldsymbol{F}d\boldsymbol{X}_2$. Consequently,

$$
\begin{aligned}
d\boldsymbol{x}_1 \cdot \boldsymbol{\sigma} d\boldsymbol{x}_2 &= (\boldsymbol{F}d\boldsymbol{X}_1) \cdot \boldsymbol{\sigma}(\boldsymbol{F}d\boldsymbol{X}_2) \\
&= d\boldsymbol{X}_1 \cdot (\boldsymbol{F}^{\mathrm{T}}\boldsymbol{\sigma}\boldsymbol{F})\, d\boldsymbol{X}_2.
\end{aligned}
\tag{5.67}
$$

This enables the time derivative to be taken independent of the material vectors, thus:

$$\frac{d}{dt}(d\boldsymbol{x}_1 \cdot \boldsymbol{\sigma} d\boldsymbol{x}_2) = d\boldsymbol{X}_1 \cdot \frac{d}{dt}(\boldsymbol{F}^{\mathrm{T}}\boldsymbol{\sigma}\boldsymbol{F})\, d\boldsymbol{X}_2. \tag{5.68}$$

Applying the chain rule to the time derivative in the above equation and noting from Equation (4.20) that $\dot{\boldsymbol{F}} = \boldsymbol{l}\boldsymbol{F}$ yields

$$\begin{aligned}\frac{d}{dt}(\boldsymbol{F}^{\mathrm{T}}\boldsymbol{\sigma}\boldsymbol{F}) &= \dot{\boldsymbol{F}}^{\mathrm{T}}\boldsymbol{\sigma}\boldsymbol{F} + \boldsymbol{F}^{\mathrm{T}}\dot{\boldsymbol{\sigma}}\boldsymbol{F} + \boldsymbol{F}^{\mathrm{T}}\boldsymbol{\sigma}\dot{\boldsymbol{F}} \\ &= \boldsymbol{F}^{\mathrm{T}}\boldsymbol{l}^{\mathrm{T}}\boldsymbol{\sigma}\boldsymbol{F} + \boldsymbol{F}^{\mathrm{T}}\dot{\boldsymbol{\sigma}}\boldsymbol{F} + \boldsymbol{F}^{\mathrm{T}}\boldsymbol{\sigma}\boldsymbol{l}\boldsymbol{F} \\ &= \boldsymbol{F}^{\mathrm{T}}(\dot{\boldsymbol{\sigma}} + \boldsymbol{\sigma}\boldsymbol{l} + \boldsymbol{l}^{\mathrm{T}}\boldsymbol{\sigma})\boldsymbol{F} \qquad (5.69) \\ &= \boldsymbol{F}^{\mathrm{T}}\boldsymbol{\sigma}^{\diamond}\boldsymbol{F}. \qquad (5.70)\end{aligned}$$

Substitute the above equation into Equation (5.68) to give

$$\begin{aligned}\frac{d}{dt}(d\boldsymbol{x}_1 \cdot \boldsymbol{\sigma} d\boldsymbol{x}_2) &= (\boldsymbol{F}d\boldsymbol{X}_1) \cdot \boldsymbol{\sigma}^{\diamond}(\boldsymbol{F}d\boldsymbol{X}_2) \\ &= d\boldsymbol{x}_1 \cdot \boldsymbol{\sigma}^{\diamond} d\boldsymbol{x}_2. \qquad (5.71)\end{aligned}$$

CHAPTER SIX

HYPERELASTICITY

Equation summary

First tensor invariant [2.47]

$$I_{\boldsymbol{S}} = \mathrm{tr}\boldsymbol{S} = \sum_{i=1}^{3} S_{ii}. \tag{6.1}$$

Pull back of the rate of deformation [4.100]

$$\dot{\boldsymbol{E}} = \boldsymbol{\phi}_*^{-1}[\boldsymbol{d}] = \boldsymbol{F}^T\boldsymbol{d}\boldsymbol{F}. \tag{6.2}$$

Constitutive equation for second Piola–Kirchhoff stress [6.7b]

$$\boldsymbol{S} = 2\frac{\partial \Psi}{\partial \boldsymbol{C}} = \frac{\partial \Psi}{\partial \boldsymbol{E}}. \tag{6.3}$$

Material elasticity tensor [6.11]

$$\boldsymbol{\mathcal{C}} = \frac{\partial \boldsymbol{S}}{\partial \boldsymbol{E}} = 2\frac{\partial \boldsymbol{S}}{\partial \boldsymbol{C}}. \tag{6.4}$$

Piola push forward of constitutive tensor [6.14]

$$\mathcal{c} = \sum_{\substack{i,j,k,l=1\\ I,J,K,L=1}}^{3} J^{-1}F_{iI}F_{jJ}F_{kK}F_{lL}\,\mathcal{C}_{IJKL}\,\boldsymbol{e}_i \otimes \boldsymbol{e}_j \otimes \boldsymbol{e}_k \otimes \boldsymbol{e}_l. \tag{6.5}$$

Derivatives of second and third invariants of a symmetric second-order tensor [6.19b, 6.22]

$$\frac{\partial II_{\boldsymbol{C}}}{\partial \boldsymbol{C}} = 2\boldsymbol{C}\,; \qquad \frac{\partial III_{\boldsymbol{C}}}{\partial \boldsymbol{C}} = III_{\boldsymbol{C}}\boldsymbol{C}^{-1}. \tag{6.6}$$

Push forward of fourth-order tensor $\mathcal{I}$ [6.39]

$$\imath = \phi_*[\mathcal{I}]; \quad \imath_{ijkl} = \sum_{I,J,K,L=1}^{3} F_{iI}F_{jJ}F_{kK}F_{lL}\mathcal{I}_{IJKL}$$
$$= \frac{1}{2}\left(\delta_{ik}\delta_{jl} + \delta_{il}\delta_{jk}\right). \tag{6.7}$$

where

$$\mathcal{I} = -\frac{\partial \boldsymbol{C}^{-1}}{\partial \boldsymbol{C}}; \quad \mathcal{I}_{IJKL} = -\frac{\partial (\boldsymbol{C}^{-1})_{IJ}}{\partial C_{KL}}. \tag{6.8}$$

Incompressible neo-Hookean material – hyperelastic potential [6.52]

$$\Psi(\boldsymbol{C}) = \frac{1}{2}\mu(\mathrm{tr}\boldsymbol{C} - 3). \tag{6.9}$$

Incompressible neo-Hookean material – second Piola–Kirchhoff stress tensor [6.54]

$$\boldsymbol{S} = \mu III_{\boldsymbol{C}}^{-1/3}(\boldsymbol{I} - \tfrac{1}{3}I_{\boldsymbol{C}}\boldsymbol{C}^{-1}) + pJ\boldsymbol{C}^{-1}. \tag{6.10}$$

Incompressible neo-Hookean material – Cauchy stress tensor [6.55]

$$\boldsymbol{\sigma} = \boldsymbol{\sigma}' + p\boldsymbol{I}; \quad \boldsymbol{\sigma}' = \mu J^{-5/3}(\boldsymbol{b} - \frac{1}{3}I_{\boldsymbol{b}}\boldsymbol{I}). \tag{6.11}$$

Isotropic elasticity in principal directions – Cauchy stress [6.79]

$$\boldsymbol{\sigma} = \sum_{\alpha=1}^{3} \sigma_{\alpha\alpha}\, \boldsymbol{n}_\alpha \otimes \boldsymbol{n}_\alpha; \quad \sigma_{\alpha\alpha} = \frac{1}{J}\frac{\partial \Psi}{\partial \ln \lambda_\alpha}. \tag{6.12}$$

Simple in-plane stretch-based hyperelastic material [6.118]

$$\sigma_{\alpha\alpha} = \frac{\bar{\lambda}}{j^\gamma}\ln j + \frac{2\mu}{j^\gamma}\ln \lambda_\alpha; \; J = j^\gamma; \; \bar{\lambda} = \frac{2\mu\lambda}{\lambda + 2\mu}. \tag{6.13}$$

EXAMPLE 6.1

A modified St. Venant–Kirchhoff constitutive behavior is defined by its corresponding strain energy functional Ψ as

$$\Psi(J, \boldsymbol{E}) = \frac{\kappa}{2}(\ln J)^2 + \mu II_{\boldsymbol{E}} \tag{6.14}$$

where $II_{\boldsymbol{E}} = \mathrm{tr}(\boldsymbol{E}^2)$ denotes the second invariant of the Green strain tensor $\boldsymbol{E}$, J is the Jacobian of the deformation gradient, and κ and μ are positive material constants.
(a) Obtain an expression for the second Piola–Kirchhoff stress tensor $\boldsymbol{S}$ as a function of the right Cauchy–Green strain tensor $\boldsymbol{C}$.
(b) Obtain an expression for the Kirchhoff stress tensor $\boldsymbol{\tau}$ as a function of the left Cauchy–Green strain tensor $\boldsymbol{b}$.
(c) Calculate the material elasticity tensor.

Solution

From Equation (6.3) the second Piola–Kirchhoff stress tensor is found as

$$\boldsymbol{S} = \frac{\partial \Psi}{\partial J}\frac{\partial J}{\partial \boldsymbol{E}} + \frac{\partial \Psi}{\partial II_{\boldsymbol{E}}}\frac{\partial II_{\boldsymbol{E}}}{\partial \boldsymbol{E}}. \tag{6.15}$$

Now consider the various terms in Equation (6.15). The simple derivatives in the above equation are easily found as

$$\frac{\partial \Psi}{\partial J} = \frac{\kappa}{J}\ln J\,; \qquad \frac{\partial \Psi}{\partial II_{\boldsymbol{E}}} = \mu. \tag{6.16}$$

Recalling that $\boldsymbol{E} = (\boldsymbol{C} - \boldsymbol{I})/2$ and, from the properties of the determinant, $J^2 = III_{\boldsymbol{C}}$ the derivative of J with respect to $\boldsymbol{E}$ is found with the aid of Equation (6.6) as

$$\frac{\partial J}{\partial \boldsymbol{E}} = \frac{\partial J}{\partial III_{\boldsymbol{C}}}\left(\frac{\partial III_{\boldsymbol{C}}}{\partial \boldsymbol{C}} : \frac{\partial \boldsymbol{C}}{\partial \boldsymbol{E}}\right) \tag{6.17a}$$

$$= \frac{1}{2} III_{\boldsymbol{C}}^{-\frac{1}{2}}\left(2\frac{\partial III_{\boldsymbol{C}}}{\partial \boldsymbol{C}}\right) \tag{6.17b}$$

$$= \frac{1}{2} III_{\boldsymbol{C}}^{-\frac{1}{2}}(2J^2\boldsymbol{C}^{-1}) \tag{6.17c}$$

$$= J\boldsymbol{C}^{-1}. \tag{6.17d}$$

Again using Equation (6.6) for the determination of the final term in Equation (6.15), the second Piola–Kirchhoff stress tensor emerges as a function of $\boldsymbol{C}$ as

$$\boldsymbol{S} = \kappa \ln J\boldsymbol{C}^{-1} + \mu(\boldsymbol{C} - \boldsymbol{I}). \tag{6.18}$$

(b) The Kirchhoff stress tensor is simply found by pushing forward the second Piola–Kirchhoff stress as given by Equation (5.9), which yields

$$\begin{aligned}\boldsymbol{\tau} = \boldsymbol{F}\boldsymbol{S}\boldsymbol{F}^{\mathrm{T}} &= \kappa \ln J\boldsymbol{F}\boldsymbol{C}^{-1}\boldsymbol{F}^{\mathrm{T}} + \mu\boldsymbol{F}\boldsymbol{C}\boldsymbol{F}^{\mathrm{T}} - \mu\boldsymbol{F}\boldsymbol{I}\boldsymbol{F}^{\mathrm{T}} \\ &= \kappa \ln J\boldsymbol{I} + \mu\boldsymbol{b}(\boldsymbol{b} - \boldsymbol{I}).\end{aligned} \tag{6.19}$$

(c) From Equation (6.4) the material elasticity tensor is

$$\boldsymbol{\mathcal{C}} = 2\frac{\partial \boldsymbol{S}}{\partial \boldsymbol{C}} = 2\frac{\partial}{\partial \boldsymbol{C}}\left(\kappa \ln J\boldsymbol{C}^{-1} + \mu(\boldsymbol{C} - \boldsymbol{I})\right) \tag{6.20a}$$

$$= 2\kappa\boldsymbol{C}^{-1} \otimes \frac{\partial \ln J}{\partial \boldsymbol{C}} + 2\kappa \ln J\frac{\partial \boldsymbol{C}^{-1}}{\partial \boldsymbol{C}} + 2\mu\frac{\partial \boldsymbol{C}}{\partial \boldsymbol{C}} \tag{6.20b}$$

$$= 2\kappa\boldsymbol{C}^{-1} \otimes \frac{\partial \ln J}{\partial J}\frac{\partial J}{\partial \boldsymbol{C}} - 2\kappa \ln J\boldsymbol{\mathcal{I}} + 2\mu\boldsymbol{\imath} \tag{6.20c}$$

$$= 2\kappa\boldsymbol{C}^{-1} \otimes \frac{1}{J}\frac{J}{2}\boldsymbol{C}^{-1} - 2\kappa \ln J\boldsymbol{\mathcal{I}} + 2\mu\boldsymbol{\imath} \tag{6.20d}$$

$$= \kappa\boldsymbol{C}^{-1} \otimes \boldsymbol{C}^{-1} - 2\kappa \ln J\boldsymbol{\mathcal{I}} + 2\mu\boldsymbol{\imath}. \tag{6.20e}$$

EXAMPLE 6.2: Textbook Exercise 6.1

In a plane stress situation, the right Cauchy–Green tensor $\boldsymbol{C}$ is

$$\boldsymbol{C} = \begin{bmatrix} C_{11} & C_{12} & 0 \\ C_{21} & C_{22} & 0 \\ 0 & 0 & C_{33} \end{bmatrix}; \quad C_{33} = \frac{h^2}{H^2},$$

where H and h are the initial and current thickness respectively. Show that incompressibility implies

$$C_{33} = III_{\overline{\boldsymbol{C}}}^{-1}\ ; \ (\boldsymbol{C}^{-1})_{33} = III_{\overline{\boldsymbol{C}}}\ ; \ \overline{\boldsymbol{C}} = \begin{bmatrix} C_{11} & C_{12} \\ C_{21} & C_{22} \end{bmatrix},$$

where $III_{\overline{\boldsymbol{C}}} = \det \overline{\boldsymbol{C}}$. Using these equations, show that for an incompressible neo-Hookean material the plane stress condition $S_{33} = 0$ enables the pressure in Equation (6.10) to be explicitly evaluated as

$$p = \tfrac{1}{3}\mu(I_{\overline{\boldsymbol{C}}} - 2III_{\overline{\boldsymbol{C}}}^{-1}),$$

and therefore the in-plane components of the second Piola–Kirchhoff and Cauchy tensors are

$$\overline{\boldsymbol{S}} = \mu\left(\bar{\boldsymbol{I}} - III_{\overline{\boldsymbol{C}}}^{-1}\,\overline{\boldsymbol{C}}^{-1}\right);$$
$$\bar{\boldsymbol{\sigma}} = \mu(\bar{\boldsymbol{b}} - III_{\bar{\boldsymbol{b}}}^{-1}\bar{\boldsymbol{I}}),$$

where the overline indicates the 2×2 in-plane components of a tensor.

Solution

The determinant of $\boldsymbol{C}$ is easily calculated as

$$\begin{aligned}\det \boldsymbol{C} &= C_{11}C_{22}C_{33} - C_{21}C_{12}C_{33}\\ &= C_{33}\det \overline{\boldsymbol{C}}\\ &= C_{33}III_{\overline{\boldsymbol{C}}}.\end{aligned} \tag{6.21}$$

Incompressibility implies that $\det \boldsymbol{C} = 1$, hence

$$C_{33} = III_{\overline{\boldsymbol{C}}}^{-1}\,; \quad C_{33}^{-1} = III_{\overline{\boldsymbol{C}}}. \tag{6.22}$$

Using Equation (6.10) the component S_{33} is found to be

$$S_{33} = \mu III_{\boldsymbol{C}}^{-1/3}(1 - \tfrac{1}{3}I_{\boldsymbol{C}}\,C_{33}^{-1}) + pJ\,C_{33}^{-1} = 0. \tag{6.23}$$

If the material is incompressible, $J = 1$, $III_{\boldsymbol{C}} = 1$, and using the second part of Equation $(6.22)_2$ yields

$$\begin{aligned}p &= -\mu(C_{33} - \tfrac{1}{3}I_{\boldsymbol{C}})\\ &= \mu(\tfrac{1}{3}I_{\boldsymbol{C}} - C_{33})\\ &= \mu(\tfrac{1}{3}I_{\overline{\boldsymbol{C}}} + \tfrac{1}{3}C_{33} - C_{33})\\ &= \tfrac{1}{3}\mu(I_{\overline{\boldsymbol{C}}} - 2C_{33})\\ &= \tfrac{1}{3}\mu(I_{\overline{\boldsymbol{C}}} - 2III_{\overline{\boldsymbol{C}}}^{-1}).\end{aligned} \tag{6.24}$$

Substituting into Equation (6.10) yields the 2×2 components of $\boldsymbol{S}$:

$$\begin{aligned}\overline{\boldsymbol{S}} &= \mu\left(\bar{\boldsymbol{I}} - \tfrac{1}{3}(C_{33} + I_{\overline{\boldsymbol{C}}})\overline{\boldsymbol{C}}^{-1}\right) + \tfrac{1}{3}\mu(I_{\overline{\boldsymbol{C}}} - 2III_{\overline{\boldsymbol{C}}}^{-1})\overline{\boldsymbol{C}}^{-1}\\ &= \mu\left(\bar{\boldsymbol{I}} - III_{\overline{\boldsymbol{C}}}^{-1}\overline{\boldsymbol{C}}^{-1}\right).\end{aligned} \tag{6.25}$$

Pushing forward using $\bar{\sigma} = J^{-1}\overline{F}\,\overline{S}\,\overline{F}^{\mathrm{T}}$, see Equation (6.11), and noting that $J = 1$ gives

$$\begin{aligned}\bar{\sigma} &= \mu J\overline{F}(\bar{I} - III_{\overline{C}}^{-1}\overline{C}^{-1})\overline{F}^{\mathrm{T}} \\ &= \mu(\overline{F}\bar{I}\overline{F}^{\mathrm{T}} - III_{\overline{C}}^{-1}\overline{F}(\overline{C}^{-1})\overline{F}^{\mathrm{T}}) \\ &= \mu(\bar{b} - III_{\bar{b}}^{-1}\overline{F}(\overline{F}^{-1}\overline{F}^{-T})\overline{F}^{\mathrm{T}}) \\ &= \mu(\bar{b} - III_{\bar{b}}^{-1}\bar{I}). \end{aligned} \tag{6.26}$$

EXAMPLE 6.3: Textbook Exercise 6.2

Show that the equations in Example 6.2 can also be derived by imposing the condition $C_{33} = III_{\overline{C}}^{-1}$ in the neo-Hookean elastic function Ψ to give

$$\Psi(\overline{C}) = \tfrac{1}{2}\mu(I_{\overline{C}} + III_{\overline{C}}^{-1} - 3),$$

from which $\overline{S}$ is obtained by differentiation with respect to the in-plane tensor $\overline{C}$. Finally, prove that the Lagrangian and Eulerian in-plane elasticity tensors are

$$\overline{\mathcal{C}} = 2\mu III_{\overline{C}}^{-1}(\overline{C}^{-1} \otimes \overline{C}^{-1} + \overline{\mathcal{I}});$$

$$\bar{c} = 2\mu III_{\bar{b}}^{-1}(\bar{I} \otimes \bar{I} + \bar{\imath}).$$

Solution

Recall the definition of the incompressible hyperelastic potential $\Psi(C)$ given by Equation (6.9) together with the trace definition given by Equation (6.1):

$$\Psi(C) = \tfrac{1}{2}\mu(\mathrm{tr}\,C - 3); \qquad \mathrm{tr}\,C = I_C, \tag{6.27}$$

where, using Equation (6.22), $I_C = I_{\overline{C}} + C_{33} = I_{\overline{C}} + III_{\overline{C}}^{-1}$. Equation (6.27) can now be written as a function of $\overline{C}$ thus:

$$\Psi(\overline{C}) = \tfrac{1}{2}\mu(I_{\overline{C}} + III_{\overline{C}}^{-1} - 3). \tag{6.28}$$

Differentiating with respect to $\overline{\boldsymbol{C}}$ allows the second Piola–Kirchhoff stress tensor $\overline{\boldsymbol{S}}$ to be evaluated as

$$\begin{aligned}
\overline{\boldsymbol{S}} &= 2\frac{\partial \Psi(\overline{\boldsymbol{C}})}{\partial \overline{\boldsymbol{C}}} \\
&= \mu \frac{\partial I_{\overline{\boldsymbol{C}}}}{\partial \overline{\boldsymbol{C}}} + \mu \frac{\partial III_{\overline{\boldsymbol{C}}}^{-1}}{\partial \overline{\boldsymbol{C}}} \\
&= \mu \bar{\boldsymbol{I}} - \mu \frac{1}{III_{\overline{\boldsymbol{C}}}^{-2}} \frac{\partial III_{\overline{\boldsymbol{C}}}}{\partial \overline{\boldsymbol{C}}} \\
&= \mu \bar{\boldsymbol{I}} - \mu III_{\overline{\boldsymbol{C}}}^{-1} \overline{\boldsymbol{C}}^{-1}.
\end{aligned} \tag{6.29}$$

Differentiating the above equation yields the Lagrangian elasticity tensor:

$$\begin{aligned}
\overline{\boldsymbol{\mathcal{C}}} &= 2\frac{\partial \overline{\boldsymbol{S}}}{\partial \overline{\boldsymbol{C}}} \\
&= -2\mu \overline{\boldsymbol{C}}^{-1} \otimes \frac{\partial III_{\overline{\boldsymbol{C}}}^{-1}}{\partial \overline{\boldsymbol{C}}} - 2\mu III_{\overline{\boldsymbol{C}}}^{-1} \frac{\partial \overline{\boldsymbol{C}}^{-1}}{\partial \overline{\boldsymbol{C}}} \\
&= 2\mu III_{\overline{\boldsymbol{C}}}^{-1} \overline{\boldsymbol{C}}^{-1} \otimes \overline{\boldsymbol{C}}^{-1} + 2\mu III_{\overline{\boldsymbol{C}}}^{-1} \overline{\boldsymbol{\mathcal{I}}}.
\end{aligned} \tag{6.30}$$

Noting the derivation of Equation (6.26) and Equation (6.7), pushing forward the above equation to the current configuration easily gives the spatial constitutive tensor $\bar{\boldsymbol{c}}$ as

$$\bar{\boldsymbol{c}} = \phi_*[\overline{\boldsymbol{\mathcal{C}}}] = 2\mu III_{\bar{\boldsymbol{b}}}^{-1} \phi_*[\boldsymbol{C}^{-1} \otimes \boldsymbol{C}^{-1}] + 2\mu III_{\bar{\boldsymbol{b}}}^{-1} \phi_*[\overline{\boldsymbol{\mathcal{I}}}], \tag{6.31}$$

where noting that $\boldsymbol{C}^{-1} = \boldsymbol{F}^{-1}\boldsymbol{F}^{-\mathrm{T}}$ and employing indicial notation yields

$$\phi_*[\boldsymbol{C}^{-1} \otimes \boldsymbol{C}^{-1}]_{ijkl}$$

$$= \sum_{I,J,K,L=1}^{2} F_{iI} F_{jJ} F_{kK} F_{lL} [\boldsymbol{C}^{-1}]_{IJ} [\boldsymbol{C}^{-1}]_{KL} \tag{6.32a}$$

$$= \sum_{I,J,K,L,m,n=1}^{2} \left[F_{iI} F_{jJ} F_{Im}^{-1} F_{Jm}^{-1}\right] \left[F_{kK} F_{lL} F_{Kn}^{-1} F_{Ln}^{-1}\right] \tag{6.32b}$$

$$= \sum_{m,n=1}^{2} [\delta_{im}\delta_{jm}] [\delta_{kn}\delta_{ln}] \tag{6.32c}$$

$$= \delta_{ij}\delta_{kl}. \tag{6.32d}$$

Similarly the push forward of $\overline{\mathcal{I}}$ is obtained from Equation (6.7) as

$$\phi_*[\overline{\mathcal{I}}]_{ijkl} = \frac{1}{2}(\delta_{ik}\delta_{jl} + \delta_{il}\delta_{jk}). \tag{6.33}$$

Combining the above expressions gives

$$\bar{\mathcal{c}} = 2\mu III_{\bar{b}}^{-1}(\bar{\boldsymbol{I}} \otimes \bar{\boldsymbol{I}} + \dot{\iota}). \tag{6.34}$$

EXAMPLE 6.4: Textbook Exercise 6.3

Using the pull back–push forward relationships between $\dot{\boldsymbol{E}}$ and $\boldsymbol{d}$ and between $\boldsymbol{\mathcal{C}}$ and $\mathcal{c}$, show that

$$\dot{\boldsymbol{E}} : \boldsymbol{\mathcal{C}} : \dot{\boldsymbol{E}} = J\boldsymbol{d} : \mathcal{c} : \boldsymbol{d}$$

for any arbitrary motion.

Solution

Recall Equation (6.2) giving the pull back of the rate of deformation $\boldsymbol{d}$ to the rate of Green's strain $\dot{\boldsymbol{E}}$:

$$\dot{\boldsymbol{E}} = \boldsymbol{F}^{\mathrm{T}}\boldsymbol{d}\boldsymbol{F}. \tag{6.35}$$

Consequently

$$\begin{aligned}
\dot{\boldsymbol{E}} : \boldsymbol{\mathcal{C}} : \dot{\boldsymbol{E}} &= (\boldsymbol{F}^{\mathrm{T}}\boldsymbol{d}\boldsymbol{F}) : \boldsymbol{\mathcal{C}} : (\boldsymbol{F}^{\mathrm{T}}\boldsymbol{d}\boldsymbol{F}) \\
&= \sum_{i,j,k,l=1}^{3} \sum_{I,J,K,L=1}^{3} F_{iI} d_{ij} F_{jJ} \mathcal{C}_{IJKL} F_{kK} d_{kl} F_{lL} \\
&= \sum_{i,j,k,l=1}^{3} \sum_{I,J,K,L=1}^{3} d_{ij} (F_{iI} F_{jJ} \mathcal{C}_{IJKL} F_{kK} F_{lL}) d_{kl} \\
&= \sum_{i,j,k,l=1}^{3} d_{ij} (J \mathcal{c}_{ijkl}) d_{kl} \\
&= J\boldsymbol{d} : \mathcal{c} : \boldsymbol{d}.
\end{aligned} \tag{6.36}$$

EXAMPLE 6.5: Textbook Exercise 6.4

Using the simple stretch-based hyperelastic equations discussed in textbook Section 6.6.7, show that the principal stresses for a simple shear test are

$$\sigma_{11} = -\sigma_{22} = 2\mu \sinh^{-1} \tfrac{\gamma}{2}.$$

Find the Cartesian stress components.

Solution

From Example 6.4, textbook page 165, the left Cauchy–Green tensor for simple shear is

$$\boldsymbol{b} = \begin{bmatrix} 1+\gamma^2 & \gamma \\ \gamma & 1 \end{bmatrix}, \tag{6.37}$$

where only two-dimensional components have been considered. The eigenvalues of $\boldsymbol{b}$ are the squared principal stretches λ_α, which satisfy the equation

$$\det \begin{bmatrix} (1+\gamma^2) - \lambda^2 & \gamma \\ \gamma & 1-\lambda^2 \end{bmatrix} = 0, \tag{6.38}$$

or

$$\left[(1+\gamma^2) - \lambda^2\right](1-\lambda^2) - \gamma^2 = 0 \tag{6.39}$$

$$(1-\lambda^2)(1-\lambda^2) = \gamma^2\lambda^2, \tag{6.40}$$

hence

$$(1-\lambda^2) = \pm\lambda\gamma \tag{6.41}$$

$$\lambda^2 \pm \lambda\gamma - 1 = 0, \tag{6.42}$$

which gives two positive values for λ:

$$\lambda_1 = \frac{\gamma}{2} + \sqrt{\left(\frac{\gamma}{2}\right)^2 + 1} \quad ; \quad \lambda_2 = -\frac{\gamma}{2} + \sqrt{\left(\frac{\gamma}{2}\right)^2 + 1}. \tag{6.43}$$

Using the simple stretch-based hyperelastic material given by Equation (6.13) and noting $J = 1$ gives

$$\sigma_{11} = 2\mu \ln\left(\frac{\gamma}{2} + \sqrt{\left(\frac{\gamma}{2}\right)^2 + 1}\right) = 2\mu \sinh^{-1}\left(\frac{\gamma}{2}\right), \tag{6.44}$$

$$\sigma_{22} = 2\mu \ln\left(-\frac{\gamma}{2} + \sqrt{\left(\frac{\gamma}{2}\right)^2 + 1}\right) = 2\mu \sinh^{-1}\left(\frac{-\gamma}{2}\right)$$
$$= -2\mu \sinh^{-1}\left(\frac{\gamma}{2}\right). \tag{6.45}$$

From Equation (6.12) the Cartesian components of the Cauchy stress are found, using the principal directions (eigenvectors) $\boldsymbol{n}_\alpha$, to be

$$\boldsymbol{\sigma} = \sum_{\alpha=1}^{2} \sigma_{\alpha\alpha}\, \boldsymbol{n}_\alpha \otimes \boldsymbol{n}_\alpha, \tag{6.46}$$

where the $\boldsymbol{n}_\alpha$ satisfy the equations

$$\begin{bmatrix} (1+\gamma^2) - \lambda^2 & \gamma \\ \gamma & 1 - \lambda^2 \end{bmatrix} \begin{bmatrix} n_\alpha^1 \\ n_\alpha^2 \end{bmatrix} = \begin{bmatrix} 0 \\ 0 \end{bmatrix}. \tag{6.47}$$

Substituting for λ_1 from Equation (6.43) gives

$$1 - \lambda_1^2 = -\frac{\gamma^2}{2} - \gamma\sqrt{\left(\frac{\gamma}{2}\right)^2 + 1}. \tag{6.48}$$

For small values of γ, $(1 - \lambda_1^2) + \gamma^2 \approx -\gamma$ which upon substitution into Equation (6.47) yields $n_1^1 = n_1^2$ from which the unit principal direction $\boldsymbol{n}_1$ is determined to be

$$\boldsymbol{n}_1 = \begin{bmatrix} \frac{1}{\sqrt{2}} \\ \frac{1}{\sqrt{2}} \end{bmatrix}. \tag{6.49}$$

Orthogonality then gives

$$\boldsymbol{n}_2 = \begin{bmatrix} -\frac{1}{\sqrt{2}} \\ \frac{1}{\sqrt{2}} \end{bmatrix} \tag{6.50}$$

and using Equations (6.44), (6.45), and (6.46) yields

$$\begin{aligned}\boldsymbol{\sigma} &= 2\mu \sinh^{-1}\left(\frac{\gamma}{2}\right)\frac{1}{2}\begin{bmatrix}1\\1\end{bmatrix} \otimes \begin{bmatrix}1 & 1\end{bmatrix}\\ &\quad -2\mu \sinh^{-1}\left(\frac{\gamma}{2}\right)\frac{1}{2}\begin{bmatrix}-1\\1\end{bmatrix} \otimes \begin{bmatrix}-1 & 1\end{bmatrix}\\ &= 2\mu \sinh^{-1}\left(\frac{\gamma}{2}\right)\begin{bmatrix}0 & 1\\1 & 0\end{bmatrix}.\end{aligned} \tag{6.51}$$

For larger values of γ the calculation proceeds in a similar fashion, but the algebra becomes more laborious.

EXAMPLE 6.6: Textbook Exercise 6.5

A general type of incompressible hyperelastic material proposed by Ogden is defined by the following strain energy function:

$$\Psi = \sum_{p=1}^{N} \frac{\mu_p}{\alpha_p}\left(\lambda_1^{\alpha_p} + \lambda_2^{\alpha_p} + \lambda_3^{\alpha_p} - 3\right).$$

Derive the homogeneous counterpart of this functional. Obtain expressions for the principal components of the deviatoric stresses and elasticity tensor.

Solution

Recalling the discussion on textbook page 168, the homogeneous counterpart of Ψ is obtained by replacing λ_α by

$$\hat{\lambda}_\alpha = J^{-1/3}\lambda_\alpha \quad ; \quad J = \lambda_1\lambda_2\lambda_3 \tag{6.52}$$

to give

$$\begin{aligned}\hat{\Psi} &= \sum_{p=1}^{N} \frac{\mu_p}{\alpha_p}\left(\hat{\lambda}_1^{\alpha_p} + \hat{\lambda}_2^{\alpha_p} + \hat{\lambda}_3^{\alpha_p} - 3\right)\\ &= \sum_{p=1}^{N} \frac{\mu_p}{\alpha_p}\left[J^{-\alpha_p/3}\left(\lambda_1^{\alpha_p} + \lambda_2^{\alpha_p} + \lambda_3^{\alpha_p} - 3\right)\right].\end{aligned} \tag{6.53}$$

Employing Equation (6.12) the principal Cauchy stresses are found to be

$$\sigma'_\beta = \frac{1}{J}\frac{\partial \hat{\Psi}}{\partial \ln \lambda_\beta} = \frac{\lambda_\beta}{J}\frac{\partial \hat{\Psi}}{\partial \lambda_\beta}, \tag{6.54}$$

where β is used to differentiate the principal direction subscript from the material coefficient superscript α_p.

Differentiating Equation (6.53) gives

$$\begin{aligned}
\sigma'_\beta &= \frac{\lambda_\beta}{J}\left[\sum_{p=1}^{N} J^{-\alpha_p/3}\,\frac{\mu_p}{\alpha_p}\alpha_p\,\lambda_\beta^{(\alpha_p-1)}\right.\\
&\qquad \left. -\frac{\mu_p}{\alpha_p}\frac{\alpha_p}{3}\,J^{(-\alpha_p/3-1)}\,\frac{J}{\lambda_\beta}\left(\lambda_1^{\alpha_p}+\lambda_2^{\alpha_p}+\lambda_3^{\alpha_p}\right)\right]\\
&= \frac{1}{J}\left[\sum_{p=1}^{N} J^{-\alpha_p/3}\,\mu_p\,\lambda_p^{\alpha_p} - J^{-\alpha_p/3}\,\frac{\mu_p}{3}\left(\lambda_1^{\alpha_p}+\lambda_2^{\alpha_p}+\lambda_3^{\alpha_p}\right)\right]\\
&= \frac{1}{J}\sum_{p=1}^{N}\mu_p\left[\hat{\lambda}_\beta^{\alpha_p} - \frac{1}{3}\left(\hat{\lambda}_1^{\alpha_p}+\hat{\lambda}_2^{\alpha_p}+\hat{\lambda}_3^{\alpha_p}\right)\right].
\end{aligned} \tag{6.55}$$

It is simple to show that $\boldsymbol{\sigma}'$ meets the requirement that $\mathrm{tr}\boldsymbol{\sigma}' = \sigma'_1 + \sigma'_2 + \sigma'_3 = 0$, see textbook page 151.

The coefficients of the spatial elasticity tensor given in the first term of textbook Equation (6.90) are obtained by differentiation as

$$\begin{aligned}
&\frac{1}{J}\frac{\partial^2 \hat{\Psi}}{\partial \ln \lambda_\beta \partial \ln \lambda_\gamma}\\
&\quad = \frac{1}{J}\frac{\partial J\sigma'_\beta}{\partial \ln \lambda_\gamma} = \frac{1}{J}\lambda_\gamma\frac{\partial J\sigma'_\beta}{\partial \lambda_\gamma}\\
&\quad = \frac{\lambda_\gamma}{J}\frac{\partial}{\partial \lambda_\gamma}\left[\sum_{p=1}^{N} J^{-\alpha_p/3}\,\mu_p\left(\lambda_\beta^{\alpha_p} - \frac{1}{3}\sum_{\delta=1}^{N}\lambda_\delta^{\alpha_p}\right)\right]
\end{aligned}$$

$$= \frac{\lambda_\gamma}{J}\left[\sum_{p=1}^{N} J^{-\alpha_p/3}\,\mu_p\left(\lambda_\gamma^{(\alpha_p-1)}\alpha_p - \frac{1}{3}\,\alpha_p\lambda_\gamma^{(\alpha_p-1)}\right)\right.$$

$$\left. - \sum_{p=1}^{N}\frac{\alpha_p}{3}\,J^{(-\alpha_p/3-1)}\,\frac{J}{\lambda_\gamma}\,\mu_p\left(\lambda_\beta^{\alpha_p} - \frac{1}{3}\sum_{\delta=1}^{N}\lambda_\delta^{\alpha_p}\right)\right]$$

$$= \frac{1}{J}\left[\sum_{p=1}^{N}\alpha_p\,\mu_p\big(\hat{\lambda}_\gamma^{\alpha_p} - \frac{1}{3}\,\hat{\lambda}_\beta^{\alpha_p} - \frac{1}{3}\,\hat{\lambda}_\gamma^{\alpha_p} + \frac{1}{9}\sum_{\delta=1}^{N}\hat{\lambda}_\delta^{\alpha_p}\big)\right]. \tag{6.56}$$

CHAPTER SEVEN

LARGE ELASTO-PLASTIC DEFORMATIONS

This chapter considers a few examples which build on the elasto-plastic formulations given in the corresponding textbook chapter.

Equation summary

Deformation gradient in principal directions [4.43]

$$\boldsymbol{F} = \sum_{\alpha=1}^{3} \lambda_\alpha \, \boldsymbol{n}_\alpha \otimes \boldsymbol{N}_\alpha. \tag{7.1}$$

Multiplicative decomposition [7.2]

$$\boldsymbol{F} = \boldsymbol{F}_e \boldsymbol{F}_p. \tag{7.2}$$

Derivative of left Cauchy–Green tensor [7.12]

$$\dot{\boldsymbol{b}}_e = \frac{d}{dt}\boldsymbol{b}_e\left(\boldsymbol{F}(t), \boldsymbol{C}_p(t)\right) = \left.\frac{d\boldsymbol{b}_e}{dt}\right|_{\boldsymbol{C}_p=\text{const}} + \left.\frac{d\boldsymbol{b}_e}{dt}\right|_{\boldsymbol{F}=\text{const}}. \tag{7.3}$$

Decomposition of total rate of work [7.13]

$$\dot{w} = \dot{w}_e + \dot{w}_p. \tag{7.4}$$

Elastic velocity gradient [7.16]

$$\boldsymbol{l}_e = \dot{\boldsymbol{F}}_e \boldsymbol{F}_e^{-1}. \tag{7.5}$$

Flow rule [7.22]

$$\boldsymbol{l}_p = -\frac{1}{2}\left.\frac{d\boldsymbol{b}_e}{dt}\right|_{\boldsymbol{F}=\text{const}} \boldsymbol{b}_e^{-1} = \dot{\gamma}\frac{\partial f(\boldsymbol{\tau}, \bar{\varepsilon}_p)}{\partial \boldsymbol{\tau}}. \tag{7.6}$$

Total work rate per unit initial volume [7.23]

$$\boldsymbol{\tau} : \boldsymbol{l} = \boldsymbol{\tau} : \boldsymbol{l}_e + \boldsymbol{\tau} : \boldsymbol{l}_p. \tag{7.7}$$

Flow rule in principal directions [7.39]

$$-\, l_{p,\alpha\alpha} = \left.\frac{d\varepsilon_{e,\alpha}}{dt}\right|_{\boldsymbol{F}=\text{const}} = -\,\dot{\gamma}\frac{\partial f(\tau_{\alpha\alpha}, \bar{\varepsilon}_p)}{\partial \tau_{\alpha\alpha}}; \quad \varepsilon_{e,\alpha} = \ln \lambda_{e,\alpha}. \tag{7.8}$$

EXAMPLE 7.1: Textbook Exercise 7.1

Using the multiplicative decomposition $\boldsymbol{F} = \boldsymbol{F}_e\boldsymbol{F}_p$ and the expressions $\boldsymbol{l} = \dot{\boldsymbol{F}}\boldsymbol{F}^{-1}$ and $\boldsymbol{l}_e = \dot{\boldsymbol{F}}_e\boldsymbol{F}_e^{-1}$, show that the plastic rate of deformation $\boldsymbol{l}_p$ can be obtained as

$$\boldsymbol{l}_p = \boldsymbol{F}_e\dot{\boldsymbol{F}}_p\boldsymbol{F}_p^{-1}\boldsymbol{F}_e^{-1}.$$

Solution

First note from Equation (7.7) that

$$\boldsymbol{\tau} : \boldsymbol{l}_p = \boldsymbol{\tau} : \boldsymbol{l} - \boldsymbol{\tau} : \boldsymbol{l}_e, \tag{7.9}$$

from which it is evident that $\boldsymbol{l}_p = \boldsymbol{l} - \boldsymbol{l}_e$. Using Equations (4.20), (7.2), and (7.5),

$$\begin{aligned}
\boldsymbol{l}_p &= \boldsymbol{l} - \boldsymbol{l}_e \\
&= \dot{\boldsymbol{F}}\boldsymbol{F}^{-1} - \dot{\boldsymbol{F}}_e\boldsymbol{F}_e^{-1} \\
&= (\dot{\boldsymbol{F}}_e\boldsymbol{F}_p + \boldsymbol{F}_e\dot{\boldsymbol{F}}_p)(\boldsymbol{F}_p^{-1}\boldsymbol{F}_e^{-1}) - \dot{\boldsymbol{F}}_e\boldsymbol{F}_e^{-1} \\
&= \dot{\boldsymbol{F}}_e\boldsymbol{F}_e^{-1} + \boldsymbol{F}_e\dot{\boldsymbol{F}}_p\boldsymbol{F}_p^{-1}\boldsymbol{F}_e^{-1} - \dot{\boldsymbol{F}}_e\boldsymbol{F}_e^{-1} \\
&= \boldsymbol{F}_e\dot{\boldsymbol{F}}_p\boldsymbol{F}_p^{-1}\boldsymbol{F}_e^{-1}.
\end{aligned} \tag{7.10}$$

EXAMPLE 7.2: Textbook Exercise 7.2

Starting from the expression $\boldsymbol{F}_e = \boldsymbol{F}_e(\boldsymbol{F}, \boldsymbol{F}_p)$ and using a decomposition similar to that shown in Equation (7.3), show that

$$\boldsymbol{l}_p = -\left.\frac{d\boldsymbol{F}_e}{dt}\right|_{\boldsymbol{F}=\text{const}} \boldsymbol{F}_e^{-1}.$$

Solution

Observing that $\boldsymbol{F}_e = \boldsymbol{F}\boldsymbol{F}_p^{-1}$ gives

$$\begin{aligned}\frac{d\boldsymbol{F}_e}{dt} &= \left.\frac{d\boldsymbol{F}_e}{dt}\right|_{\boldsymbol{F}_p=\text{ const}} + \left.\frac{d\boldsymbol{F}_e}{dt}\right|_{\boldsymbol{F}=\text{const}} \\ &= \dot{\boldsymbol{F}}\boldsymbol{F}_p^{-1} + \left.\frac{d\boldsymbol{F}_e}{dt}\right|_{\boldsymbol{F}=\text{const}}.\end{aligned} \tag{7.11}$$

Hence,

$$\begin{aligned}\left.\frac{d\boldsymbol{F}_e}{dt}\right|_{\boldsymbol{F}=\text{const}} \boldsymbol{F}_e^{-1} &= \dot{\boldsymbol{F}}_e\boldsymbol{F}_e^{-1} - \dot{\boldsymbol{F}}\boldsymbol{F}_p^{-1}\boldsymbol{F}_e^{-1} \\ &= \boldsymbol{l}_e - \boldsymbol{l} \\ &= -\boldsymbol{l}_p.\end{aligned} \tag{7.12}$$

EXAMPLE 7.3: Textbook Exercise 7.3

Use Equation (7.12) to derive the flow rule in principal directions following a procedure similar to that described in Section 7.5.

Solution

From Equation (7.6),

$$\boldsymbol{l}_p = \dot{\gamma}\frac{\partial f(\boldsymbol{\tau}, \bar{\varepsilon}_p)}{\partial \boldsymbol{\tau}}, \tag{7.13}$$

therefore

$$-\left.\frac{d\boldsymbol{F}_e}{dt}\right|_{\boldsymbol{F}=\text{const}} \boldsymbol{F}_e^{-1} = \dot{\gamma}\frac{\partial f(\boldsymbol{\tau}, \bar{\varepsilon}_p)}{\partial \boldsymbol{\tau}}. \tag{7.14}$$

In principal directions the elastic component of the deformation gradient can be expressed as

$$\boldsymbol{F}_e = \sum_{\alpha=1}^{3} \lambda_{e,\alpha}\, \boldsymbol{n}_\alpha \otimes \boldsymbol{n}^p_\alpha; \quad \boldsymbol{F}_e^{-1} = \sum_{\alpha=1}^{3} \lambda_{e,\alpha}^{-1}\, \boldsymbol{n}^p_\alpha \otimes \boldsymbol{n}_\alpha, \tag{7.15}$$

where $\boldsymbol{n}^p_\alpha$ represents the principal directions in the plastic reference state (see, for comparison, Equation (7.1)). Consequently,

$$
\begin{aligned}
-\boldsymbol{l}_p &= \left.\frac{d\boldsymbol{F}_e}{dt}\right|_{\boldsymbol{F}} \boldsymbol{F}_e^{-1} && (7.16)\\
&= \sum_{\alpha=1}^{3} \left.\frac{d\lambda_{e,\alpha}}{dt}\right|_{\boldsymbol{F}} \lambda_{e,\alpha}^{-1}\, \boldsymbol{n}_\alpha \otimes \boldsymbol{n}_\alpha\\
&+ \sum_{\alpha=1}^{3} \lambda_{e,\alpha}\, \lambda_{e,\alpha}^{-1} \left.\frac{d\boldsymbol{n}_\alpha}{dt}\right|_{\boldsymbol{F}} \otimes \boldsymbol{n}_\alpha\\
&+ \sum_{\alpha,\beta=1}^{3} \lambda_{e,\alpha}\, \lambda_{e,\beta}^{-1} \left(\left.\frac{d\boldsymbol{n}^p_\alpha}{dt}\right|_{\boldsymbol{F}} \cdot \boldsymbol{n}^p_\beta \right) \boldsymbol{n}_\alpha \otimes \boldsymbol{n}_\beta\\
&= \sum_{\alpha=1}^{3} \left.\frac{d\lambda_{e,\alpha}}{dt}\right|_{\boldsymbol{F}} \lambda_{e,\alpha}^{-1}\, \boldsymbol{n}_\alpha \otimes \boldsymbol{n}_\alpha\\
&+ \sum_{\substack{\alpha,\beta=1\\ \alpha\neq\beta}}^{3} \left(\left.\frac{d\boldsymbol{n}_\beta}{dt}\right|_{\boldsymbol{F}} \cdot \boldsymbol{n}_\alpha + \left.\frac{d\boldsymbol{n}^p_\alpha}{dt}\right|_{\boldsymbol{F}} \cdot \boldsymbol{n}^p_\beta \left(\frac{\lambda_{e,\alpha}}{\lambda_{e,\beta}} \right) \right) \boldsymbol{n}_\alpha \otimes \boldsymbol{n}_\beta.
\end{aligned}
$$

(7.17)

Using again the symmetry arguments deployed in book Section 7.5 between Equations (7.35) and (7.37) leads to

$$
\begin{aligned}
\boldsymbol{l}_p &= -\sum_{\alpha=1}^{3} \left.\frac{d\lambda_{e,\alpha}}{dt}\right|_{\boldsymbol{F}} \lambda_{e,\alpha}^{-1}\, \boldsymbol{n}_\alpha \otimes \boldsymbol{n}_\alpha\\
&= -\sum_{\alpha=1}^{3} \left.\frac{d\ln\lambda_{e,\alpha}}{dt}\right|_{\boldsymbol{F}} \boldsymbol{n}_\alpha \otimes \boldsymbol{n}_\alpha. && (7.18)
\end{aligned}
$$

Or

$$l_{p,\alpha\alpha} = -\left.\frac{d\varepsilon_{e,\alpha}}{dt}\right|_{F=\text{const}} = \dot{\gamma}\frac{\partial f(\tau_{\alpha\alpha},\bar{\varepsilon}_p)}{\partial \tau_{\alpha\alpha}}, \tag{7.19}$$

which coincides with Equation (7.8).

EXAMPLE 7.4: Textbook Exercise 7.4

Consider a material in which the internal elastic energy is expressed as $\Psi(\boldsymbol{C}, \boldsymbol{C}_p)$. Show that the plastic dissipation rate can be expressed as

$$\dot{w}_p = -\frac{\partial \Psi}{\partial \boldsymbol{C}_p} : \dot{\boldsymbol{C}}_p.$$

Starting from this expression and using the principle of maximum plastic dissipation, show that if the yield surface is defined in terms of $\boldsymbol{C}$ and $\boldsymbol{C}_p$ by $f(\boldsymbol{C}, \boldsymbol{C}_p) \leq 0$ then the flow rule becomes

$$\frac{\partial^2 \Psi}{\partial \boldsymbol{C} \partial \boldsymbol{C}_P} : \dot{\boldsymbol{C}}_p = -\gamma \frac{\partial f}{\partial \boldsymbol{C}}.$$

Solution

From the definition given by Equation (7.4),

$$\begin{aligned} \dot{w}_p &= \dot{w} - \dot{w}_e \\ &= \tfrac{1}{2}\boldsymbol{S} : \dot{\boldsymbol{C}} - \dot{\Psi}, \end{aligned} \tag{7.20}$$

where $\boldsymbol{S} = 2\frac{\partial \Psi}{\partial \boldsymbol{C}}$, and $\dot{\Psi}$ can be found as

$$\dot{\Psi} = \frac{\partial \Psi}{\partial \boldsymbol{C}} : \dot{\boldsymbol{C}} + \frac{\partial \Psi}{\partial \boldsymbol{C}_p} : \dot{\boldsymbol{C}}_p. \tag{7.21}$$

Combining these equations gives

$$\begin{aligned} \dot{w}_p &= \frac{1}{2} 2\frac{\partial \Psi}{\partial \boldsymbol{C}} : \dot{\boldsymbol{C}} - \frac{\partial \Psi}{\partial \boldsymbol{C}} : \dot{\boldsymbol{C}} - \frac{\partial \Psi}{\partial \boldsymbol{C}_p} : \dot{\boldsymbol{C}}_p \\ &= -\frac{\partial \Psi}{\partial \boldsymbol{C}_p} : \dot{\boldsymbol{C}}_p. \end{aligned} \tag{7.22}$$

Maximizing the plastic dissipation subject to the condition $f(\boldsymbol{C}, \boldsymbol{C}_p) \le 0$ is achieved by using a Lagrange multiplier γ to define the functional

$$\Pi = \dot{w}_p(\boldsymbol{C}, \boldsymbol{C}_p) + \gamma f(\boldsymbol{C}, \boldsymbol{C}_p). \tag{7.23}$$

The stationary conditions of this functional imply

$$\frac{\partial \Psi}{\partial \boldsymbol{C} \partial \boldsymbol{C}_p} : \dot{\boldsymbol{C}}_p = -\gamma \frac{\partial f}{\partial \boldsymbol{C}}. \tag{7.24}$$

CHAPTER EIGHT

LINEARIZED EQUILIBRIUM EQUATIONS

This chapter presents various examples involving linearization of terms in the virtual work expression of equilibrium. Such linearization, which employs the concept of the directional derivative, is central to the formulation of the Newton–Raphson process necessary for achieving an eventual solution to the discretized nonlinear equilibrium equations. A final example considers the linearization of the six-field Hu–Washizu variational principle.

Equation summary

Directional derivative of the deformation gradient tensor [4.70]

$$D\boldsymbol{F}[\boldsymbol{u}] = \frac{\partial \boldsymbol{u}(\boldsymbol{X})}{\partial \boldsymbol{X}} = \boldsymbol{\nabla}_0 \boldsymbol{u}. \tag{8.1}$$

Internal virtual work in terms of first Piola–Kirchhoff stress [5.33]

$$\delta W_{\text{int}}(\phi, \delta \boldsymbol{v}) = \int_V \boldsymbol{P} : \delta \dot{\boldsymbol{F}}\, dV. \tag{8.2}$$

Condition on the deviatoric component of the first Piola–Kirchhoff stress tensor [5.51b]

$$\boldsymbol{P}' : \boldsymbol{F} = 0. \tag{8.3}$$

First Piola–Kirchhoff stress tensor as a function of an elastic potential [6.5]

$$\boldsymbol{P}(\boldsymbol{F}(\boldsymbol{X}),\boldsymbol{X})=\frac{\partial\Psi(\boldsymbol{F}(\boldsymbol{X}),\boldsymbol{X})}{\partial\boldsymbol{F}}. \tag{8.4}$$

Virtual deformation gradient rate [8.7]

$$\delta\dot{\boldsymbol{F}}=\frac{\partial\delta\boldsymbol{v}}{\partial\boldsymbol{X}}=\boldsymbol{\nabla}_0\delta\boldsymbol{v}. \tag{8.5}$$

Surface normal vector [8.19]

$$\boldsymbol{n}=\frac{\frac{\partial\boldsymbol{x}}{\partial\xi}\times\frac{\partial\boldsymbol{x}}{\partial\eta}}{\left\|\frac{\partial\boldsymbol{x}}{\partial\xi}\times\frac{\partial\boldsymbol{x}}{\partial\eta}\right\|};\qquad da=\left\|\frac{\partial\boldsymbol{x}}{\partial\xi}\times\frac{\partial\boldsymbol{x}}{\partial\eta}\right\|d\xi d\eta, \tag{8.6}$$

Various forms of the directional derivative of the virtual external work pressure component [8.20]

$$\delta W^p_{\text{ext}}(\boldsymbol{\phi},\delta\boldsymbol{v})=\int_{A_\xi}p\delta\boldsymbol{v}\cdot\left(\frac{\partial\boldsymbol{x}}{\partial\xi}\times\frac{\partial\boldsymbol{x}}{\partial\eta}\right)d\xi d\eta. \tag{8.7}$$

Directional derivative of the virtual external work pressure component [8.21]

$$\begin{aligned}D\delta W^p_{\text{ext}}(\boldsymbol{\phi},\delta\boldsymbol{v})[\boldsymbol{u}]=\int_{A_\xi}p\bigg[&\frac{\partial\boldsymbol{x}}{\partial\xi}\cdot\left(\frac{\partial\boldsymbol{u}}{\partial\eta}\times\delta\boldsymbol{v}\right)\\&-\frac{\partial\boldsymbol{x}}{\partial\eta}\cdot\left(\frac{\partial\boldsymbol{u}}{\partial\xi}\times\delta\boldsymbol{v}\right)\bigg]d\xi d\eta.\end{aligned} \tag{8.8}$$

Directional derivative of the virtual external work pressure component (8.21 rewritten) [8.22]

$$\begin{aligned}D\delta W^p_{\text{ext}}(\boldsymbol{\phi},\delta\boldsymbol{v})[\boldsymbol{u}]=\int_{A_\xi}p\bigg[&\frac{\partial\boldsymbol{x}}{\partial\xi}\cdot\left(\frac{\partial\delta\boldsymbol{v}}{\partial\eta}\times\boldsymbol{u}\right)\\&-\frac{\partial\boldsymbol{x}}{\partial\eta}\cdot\left(\frac{\partial\delta\boldsymbol{v}}{\partial\xi}\times\boldsymbol{u}\right)\bigg]d\xi d\eta\\&+\oint_{\partial A_\xi}p(\boldsymbol{u}\times\delta\boldsymbol{v})\cdot\left(\nu_\eta\frac{\partial\boldsymbol{x}}{\partial\xi}-\nu_\xi\frac{\partial\boldsymbol{x}}{\partial\eta}\right)dl.\end{aligned}$$

$$\boldsymbol{\nu}=[\nu_\xi,\nu_\eta]^T\text{ vector normal to }\partial A_{\boldsymbol{\xi}}. \tag{8.9}$$

Directional derivative of the virtual external work pressure component for a closed boundary [8.23]

$$D\delta W^{p}_{\text{ext}}(\boldsymbol{\phi},\delta\boldsymbol{v})[\boldsymbol{u}] = \frac{1}{2}\int_{A_\xi} p\frac{\partial \boldsymbol{x}}{\partial \xi}\cdot\left[\left(\frac{\partial \boldsymbol{u}}{\partial \eta}\times\delta\boldsymbol{v}\right)+\left(\frac{\partial \delta\boldsymbol{v}}{\partial \eta}\times\boldsymbol{u}\right)\right]d\xi d\eta$$
$$-\frac{1}{2}\int_{A_\xi} p\frac{\partial \boldsymbol{x}}{\partial \eta}\cdot\left[\left(\frac{\partial \boldsymbol{u}}{\partial \xi}\times\delta\boldsymbol{v}\right)\right.$$
$$\left.+\left(\frac{\partial \delta\boldsymbol{v}}{\partial \xi}\times\boldsymbol{u}\right)\right]d\xi d\eta. \tag{8.10}$$

Stationary condition of the perturbed Lagrangian functional employed in the Penalty Method for Incompressibility [8.41]

$$D\Pi_P(\boldsymbol{\phi},p)[\delta p] = \int_V \delta p\left[(J-1)-\frac{p}{\kappa}\right]dV = 0. \tag{8.11}$$

Mean pressure from Mean Dilatation Procedure [8.52]

$$p(\bar{J}) = \kappa\left(\frac{v-V}{V}\right). \tag{8.12}$$

EXAMPLE 8.1: Textbook Exercise 8.1

Show that the linearized internal virtual work can also be expressed as

$$D\delta W(\boldsymbol{\phi},\delta\boldsymbol{v})[\boldsymbol{u}] = \int_V (\boldsymbol{\nabla}_0\delta\boldsymbol{v}) : \boldsymbol{\mathcal{A}} : (\boldsymbol{\nabla}_0\boldsymbol{u})\,dV; \boldsymbol{\mathcal{A}} = \frac{\partial \boldsymbol{P}}{\partial \boldsymbol{F}} - \frac{\partial^2\Psi}{\partial \boldsymbol{F}\partial \boldsymbol{F}},$$

where $\boldsymbol{P}$ is the first Piola–Kirchhoff tensor.

Solution

Taking the directional derivative of Equation (8.2) in the direction of an incremental change in displacement $\boldsymbol{u}$ and using Equation (8.5) gives

$$D\delta W_{\text{int}}(\boldsymbol{\phi},\delta\boldsymbol{v})[\boldsymbol{u}] = \int_V D\boldsymbol{P}[\boldsymbol{u}] : \boldsymbol{\nabla}_0\delta\boldsymbol{v}\,dV, \tag{8.13}$$

where it is noted that $\delta\boldsymbol{v}$ is not a function of $\boldsymbol{u}$.

Observing from Equation (8.4) that $\boldsymbol{P}$ is a function of $\boldsymbol{F}$ and recalling from Equation (8.1) that $D\boldsymbol{F}[\boldsymbol{u}] = \boldsymbol{\nabla}_0 \boldsymbol{u}$ gives

$$\begin{aligned} D\boldsymbol{P}[\boldsymbol{u}] &= \frac{\partial \boldsymbol{P}}{\partial \boldsymbol{F}} : D\boldsymbol{F}[\boldsymbol{u}] \\ &= \frac{\partial^2 \Phi}{\partial \boldsymbol{F} \partial \boldsymbol{F}} : D\boldsymbol{F}[\boldsymbol{u}] \\ &= \boldsymbol{\mathcal{A}} : \boldsymbol{\nabla}_0 \boldsymbol{u}. \end{aligned} \tag{8.14}$$

Substituting into Equation (8.13) yields

$$D\delta W_{\text{int}}(\phi, \delta \boldsymbol{v})[\boldsymbol{u}] = \int_V (\boldsymbol{\nabla}_0 \delta \boldsymbol{v}) : \boldsymbol{\mathcal{A}} : (\boldsymbol{\nabla}_0 \boldsymbol{u})\, dV. \tag{8.15}$$

EXAMPLE 8.2: Textbook Exercise 8.2

Show that for the case of uniform pressure over an enclosed fixed boundary, the external virtual work can be derived from an associated potential as $\delta W^p_{\text{ext}}(\boldsymbol{\phi}, \delta \boldsymbol{v}) = D\Pi^p_{\text{ext}}(\boldsymbol{\phi})[\delta \boldsymbol{v}]$, where

$$\Pi^p_{\text{ext}}(\boldsymbol{\phi}) = \frac{1}{3} \int_a p\, \boldsymbol{x} \cdot \boldsymbol{n}\, da.$$

Explain the physical significance of this integral.

Solution

From Equation (8.6) and noting Figure 8.1, the external pressure potential becomes

$$\Pi^p_{\text{ext}}(\boldsymbol{\phi}) = \frac{1}{3} \int_{A_\xi} p\, \boldsymbol{x} \cdot \left(\frac{\partial \boldsymbol{x}}{\partial \xi} \times \frac{\partial \boldsymbol{x}}{\partial \eta} \right) d\xi d\eta. \tag{8.16}$$

Taking the directional derivative of $\Pi^p_{\text{ext}}(\boldsymbol{\phi})$ in the direction of a virtual velocity $\delta \boldsymbol{v}$ gives

$$\begin{aligned} D\Pi^p_{\text{ext}}(\boldsymbol{\phi})[\delta \boldsymbol{v}] &= \frac{1}{3} \int_{A_\xi} p\, \delta \boldsymbol{v} \cdot \left(\frac{\partial \boldsymbol{x}}{\partial \xi} \times \frac{\partial \boldsymbol{x}}{\partial \eta} \right) d\xi d\eta \\ &\quad + \frac{1}{3} \int_{A_\xi} p\, \boldsymbol{x} \cdot \left(\frac{\partial \delta \boldsymbol{v}}{\partial \xi} \times \frac{\partial \boldsymbol{x}}{\partial \eta} \right) d\xi d\eta \\ &\quad + \frac{1}{3} \int_{A_\xi} p\, \boldsymbol{x} \cdot \left(\frac{\partial \boldsymbol{x}}{\partial \xi} \times \frac{\partial \delta \boldsymbol{v}}{\partial \eta} \right) d\xi d\eta. \end{aligned} \tag{8.17}$$

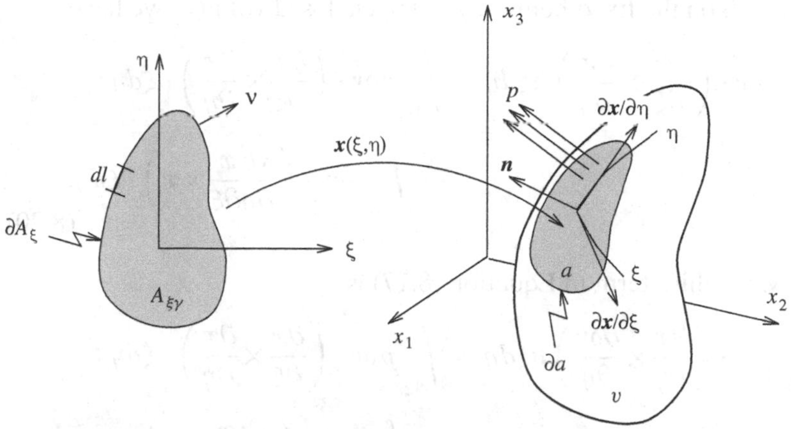

FIGURE 8.1 Uniform surface pressure

It is now possible to show that both the second and third terms in the above equation are equal to the first term for the case of an enclosed volume under uniform pressure. For example,

$$\int_{A_\xi} p\boldsymbol{x}\cdot\left(\frac{\partial\delta\boldsymbol{v}}{\partial\xi}\times\frac{\partial\boldsymbol{x}}{\partial\eta}\right)d\xi d\eta = p\int_{A_\xi}\frac{\partial\delta\boldsymbol{v}}{\partial\xi}\cdot\left(\frac{\partial\boldsymbol{x}}{\partial\eta}\times\boldsymbol{x}\right)d\xi d\eta. \quad (8.18)$$

Using the integral theorems of textbook section 2.4.2 gives

$$\begin{aligned}\int_{A_\xi} p\boldsymbol{x}\cdot\left(\frac{\partial\delta\boldsymbol{v}}{\partial\xi}\times\frac{\partial\boldsymbol{x}}{\partial\eta}\right)d\xi d\eta &= p\int_{A_\xi}\frac{\partial}{\partial\xi}\left[\delta\boldsymbol{v}\cdot\left(\frac{\partial\boldsymbol{x}}{\partial\eta}\times\boldsymbol{x}\right)\right]d\xi d\eta\\ &\quad -p\int_{A_\xi}\delta\boldsymbol{v}\cdot\left(\frac{\partial\boldsymbol{x}}{\partial\eta}\times\frac{\partial\boldsymbol{x}}{\partial\xi}\right)d\xi d\eta\\ &\quad -p\int_{A_\xi}\delta\boldsymbol{v}\cdot\left(\frac{\partial^2\boldsymbol{x}}{\partial\eta\partial\xi}\times\boldsymbol{x}\right)d\xi d\eta\\ &= p\oint_{\partial A_\xi}\delta\boldsymbol{v}\cdot\left(\frac{\partial\boldsymbol{x}}{\partial\eta}\times\boldsymbol{x}\right)d\eta\\ &\quad +\int_{A_\xi}p\,\delta\boldsymbol{v}\cdot\left(\frac{\partial\boldsymbol{x}}{\partial\xi}\times\frac{\partial\boldsymbol{x}}{\partial\eta}\right)d\xi d\eta\\ &\quad -\int_{A_\xi}p\,\delta\boldsymbol{v}\cdot\left(\frac{\partial^2\boldsymbol{x}}{\partial\eta\partial\xi}\times\boldsymbol{x}\right)d\xi d\eta.\end{aligned} \quad (8.19)$$

Since $\delta\boldsymbol{v} = \mathbf{0}$ on the fixed boundary of the enclosed volume, we have

$$\int_{A_\xi} p\,\boldsymbol{x}\cdot\left(\frac{\partial\delta\boldsymbol{v}}{\partial\xi}\times\frac{\partial\boldsymbol{x}}{\partial\eta}\right)d\xi d\eta = \int_{A_\xi} p\,\delta\boldsymbol{v}\cdot\left(\frac{\partial\boldsymbol{x}}{\partial\xi}\times\frac{\partial\boldsymbol{x}}{\partial\eta}\right)d\xi d\eta$$
$$- \int_{A_\xi} p\,\delta\boldsymbol{v}\cdot\left(\frac{\partial^2\boldsymbol{x}}{\partial\eta\partial\xi}\times\boldsymbol{x}\right)d\xi d\eta. \tag{8.20}$$

Similarly, the third term in Equation (8.17) is

$$\int_{A_\xi} p\,\boldsymbol{x}\cdot\left(\frac{\partial\boldsymbol{x}}{\partial\xi}\times\frac{\partial\delta\boldsymbol{v}}{\partial\eta}\right)d\xi d\eta = \int_{A_\xi} p\,\delta\boldsymbol{v}\cdot\left(\frac{\partial\boldsymbol{x}}{\partial\xi}\times\frac{\partial\boldsymbol{x}}{\partial\eta}\right)d\xi d\eta$$
$$+ \int_{A_\xi} p\,\delta\boldsymbol{v}\cdot\left(\frac{\partial^2\boldsymbol{x}}{\partial\xi\partial\eta}\times\boldsymbol{x}\right)d\xi d\eta. \tag{8.21}$$

Consequently,

$$\begin{aligned}
D\Pi^p_{\text{ext}}(\boldsymbol{\phi})[\delta\boldsymbol{v}] &= \frac{1}{3}\int_{A_\xi} p\,\delta\boldsymbol{v}\cdot\left(\frac{\partial\boldsymbol{x}}{\partial\xi}\times\frac{\partial\boldsymbol{x}}{\partial\eta}\right)d\xi d\eta \\
&\quad+\frac{1}{3}\int_{A_\xi} p\,\delta\boldsymbol{v}\cdot\left(\frac{\partial\boldsymbol{x}}{\partial\xi}\times\frac{\partial\boldsymbol{x}}{\partial\eta}\right)d\xi d\eta \\
&\quad+\frac{1}{3}\int_{A_\xi} p\,\delta\boldsymbol{v}\cdot\left(\frac{\partial\boldsymbol{x}}{\partial\xi}\times\frac{\partial\boldsymbol{x}}{\partial\eta}\right)d\xi d\eta \\
&= \int_{A_\xi} p\,\delta\boldsymbol{v}\cdot\left(\frac{\partial\boldsymbol{x}}{\partial\xi}\times\frac{\partial\boldsymbol{x}}{\partial\eta}\right)d\xi d\eta \\
&= \int_a p\,\delta\boldsymbol{v}\cdot\boldsymbol{n}\,da \\
&= \delta W^p_{\text{ext}}(\boldsymbol{\phi},\delta\boldsymbol{v}).
\end{aligned} \tag{8.22}$$

The term $\frac{1}{3}\int_a p\,\boldsymbol{x}\cdot\boldsymbol{n}\,da$ represents the energy stored in the gas contained by the surface at pressure p.

EXAMPLE 8.3: Textbook Exercise 8.3

Prove that for two-dimensional applications, Equation (8.10) becomes

$$D\delta W^p_{\text{ext}}(\boldsymbol{\phi},\delta\boldsymbol{v})[\boldsymbol{u}] = \frac{1}{2}\int_{L_\eta} p\boldsymbol{k}\cdot\left[\left(\frac{\partial\boldsymbol{u}}{\partial\eta}\times\delta\boldsymbol{v}\right)+\left(\frac{\partial\delta\boldsymbol{v}}{\partial\eta}\times\boldsymbol{u}\right)\right]d\eta,$$

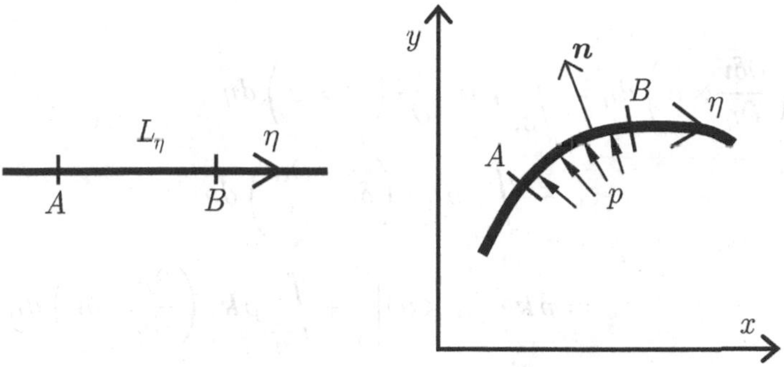

FIGURE 8.2 Example 8.3: Two-dimensional pressure linearization

where $\boldsymbol{k}$ is a unit vector normal to the two-dimensional plane and η is a parameter along the line L_η where the pressure p is applied.

Solution

Figure 8.2 is the two-dimensional equivalent of textbook Figure 8.1. The unit vector $\boldsymbol{k}$ in the z-direction is perpendicular to the x–y plane. The position vector $\boldsymbol{x}(\eta)$ is parameterized in terms of the one-dimensional coordinate η which enables the normal $\boldsymbol{n}$ to be found to be

$$\boldsymbol{n} = \frac{\boldsymbol{k}\times\dfrac{\partial \boldsymbol{x}}{\partial \eta}}{\left\|\dfrac{\partial \boldsymbol{x}}{\partial \eta}\right\|}. \tag{8.23}$$

For the two-dimensional case Equation (8.7) becomes

$$\delta W^{p}_{\text{ext}}(\boldsymbol{\phi}, \delta\boldsymbol{v}) = \int_{L_\eta} p\,\delta\boldsymbol{v}\cdot\left(\boldsymbol{k}\times\frac{\partial \boldsymbol{x}}{\partial \eta}\right) d\eta. \tag{8.24}$$

Linearizing with respect to an increment in displacements $\boldsymbol{u}$ gives

$$\begin{aligned} D\delta W^{p}_{\text{ext}}(\boldsymbol{\phi}, \delta\boldsymbol{v})[\boldsymbol{u}] &= \int_{L_\eta} p\,\delta\boldsymbol{v}\cdot\left(\boldsymbol{k}\times\frac{\partial \boldsymbol{u}}{\partial \eta}\right) d\eta \\ &= \int_{L_\eta} p\,\boldsymbol{k}\cdot\left(\frac{\partial \boldsymbol{u}}{\partial \eta}\times\delta\boldsymbol{v}\right) d\eta. \end{aligned} \tag{8.25}$$

For enclosed boundaries with uniform pressure where $\delta\boldsymbol{v}$ and $\boldsymbol{u}$ vanish at the edges, we can evaluate the right-hand side of Equation (8.25) as follows:

$$\begin{aligned}\int_{L_\eta} p\boldsymbol{k}\cdot\left(\frac{\partial\delta\boldsymbol{v}}{\partial\eta}\times\boldsymbol{u}\right)d\eta &= \int_{L_\eta} p\boldsymbol{k}\cdot\frac{\partial}{\partial\eta}\left(\delta\boldsymbol{v}\times\boldsymbol{u}\right)d\eta \\ &\quad -\int_{L_\eta} p\boldsymbol{k}\cdot\left(\delta\boldsymbol{v}\times\frac{\partial\boldsymbol{u}}{\partial\eta}\right)d\eta \\ &= p\boldsymbol{k}\cdot(\delta\boldsymbol{v}\times\boldsymbol{u})\Big|_A^B + \int_{L_\eta} p\boldsymbol{k}\cdot\left(\frac{\partial\boldsymbol{u}}{\partial\eta}\times\delta\boldsymbol{v}\right)d\eta \\ &= \int_{L_\eta} p\boldsymbol{k}\cdot\left(\frac{\partial\boldsymbol{u}}{\partial\eta}\times\delta\boldsymbol{v}\right)d\eta, \end{aligned} \tag{8.26}$$

where A and B are the left and right boundary points of L_η, where both $\boldsymbol{u}$ and $\delta\boldsymbol{v}$ are zero. Hence

$$D\delta W^p_{\rm ext}(\boldsymbol{\phi},\delta\boldsymbol{v})[\boldsymbol{u}] = \frac{1}{2}\int_{L_\eta} p\boldsymbol{k}\cdot\left[\left(\frac{\partial\boldsymbol{u}}{\partial\eta}\times\delta\boldsymbol{v}\right)+\left(\frac{\partial\delta\boldsymbol{v}}{\partial\eta}\times\boldsymbol{u}\right)\right]d\eta. \tag{8.27}$$

EXAMPLE 8.4: Textbook Exercise 8.4

Prove that by using a different cyclic permutation than that used to derive Equation (8.9), the following alternative form of Equation (8.10) can be found for the case of an enclosed fixed boundary with uniform surface pressure:

$$D\delta W^p_{\rm ext}(\boldsymbol{\phi},\delta\boldsymbol{v})[\boldsymbol{u}] = \int_{A_\xi} p\boldsymbol{x}\cdot\left[\left(\frac{\partial\delta\boldsymbol{v}}{\partial\xi}\times\frac{\partial\boldsymbol{u}}{\partial\eta}\right)-\left(\frac{\partial\delta\boldsymbol{v}}{\partial\eta}\times\frac{\partial\boldsymbol{u}}{\partial\xi}\right)\right]d\xi d\eta. \tag{8.28}$$

Solution

Starting from Equation (8.8), the process is similar to Example 8.2. Consider, for instance, the first term in Equation (8.8) (eqn. continues on next page):

$$\begin{aligned}\int_{A_\xi} p\frac{\partial\boldsymbol{x}}{\partial\xi}\cdot\left(\frac{\partial\boldsymbol{u}}{\partial\eta}\times\delta\boldsymbol{v}\right)d\xi d\eta &= \int_{A_\xi} p\frac{\partial}{\partial\xi}\left[\boldsymbol{x}\cdot\left(\frac{\partial\boldsymbol{u}}{\partial\eta}\times\delta\boldsymbol{v}\right)\right]d\xi d\eta \\ &\quad -\int_{A_\xi} p\boldsymbol{x}\cdot\left(\frac{\partial\boldsymbol{u}}{\partial\eta}\times\frac{\partial\delta\boldsymbol{v}}{\partial\xi}\right)d\xi d\eta\end{aligned}$$

$$
\begin{aligned}
&- \int_{A_\xi} p\boldsymbol{x} \cdot \left(\frac{\partial^2 \boldsymbol{u}}{\partial \xi \partial \eta} \times \delta \boldsymbol{v} \right) d\xi d\eta \\
&= \int_{A_\xi} p\boldsymbol{x} \cdot \left(\frac{\partial \delta \boldsymbol{v}}{\partial \xi} \times \frac{\partial \boldsymbol{u}}{\partial \eta} \right) d\xi d\eta \\
&- \int_{A_\xi} p\boldsymbol{x} \cdot \left(\frac{\partial^2 \boldsymbol{u}}{\partial \xi \partial \eta} \times \delta \boldsymbol{v} \right) d\xi d\eta.
\end{aligned} \tag{8.29}
$$

Similarly,

$$
\begin{aligned}
\int_{A_\xi} p \frac{\partial \boldsymbol{x}}{\partial \eta} \cdot \left(\frac{\partial \boldsymbol{u}}{\partial \xi} \times \delta \boldsymbol{v} \right) d\xi d\eta = &\int_{A_\xi} p\boldsymbol{x} \cdot \left(\frac{\partial \delta \boldsymbol{v}}{\partial \eta} \times \frac{\partial \boldsymbol{u}}{\partial \xi} \right) d\xi d\eta \\
&- \int_{A_\xi} p\boldsymbol{x} \cdot \left(\frac{\partial^2 \boldsymbol{u}}{\partial \xi \partial \eta} \times \delta \boldsymbol{v} \right) d\xi d\eta.
\end{aligned} \tag{8.30}
$$

Recalling Equation (8.8), $D\delta W^p_{\text{ext}}(\boldsymbol{\phi}, \delta\boldsymbol{v})[\boldsymbol{u}]$ is found by subtracting Equation (8.30) from Equation (8.29) to give

$$
D\delta W^p_{\text{ext}}(\boldsymbol{\phi}, \delta\boldsymbol{v})[\boldsymbol{u}] = \int_{A_\xi} p\boldsymbol{x} \cdot \left[\left(\frac{\partial \delta \boldsymbol{v}}{\partial \xi} \times \frac{\partial \boldsymbol{u}}{\partial \eta} \right) - \left(\frac{\partial \delta \boldsymbol{v}}{\partial \eta} \times \frac{\partial \boldsymbol{u}}{\partial \xi} \right) \right] d\xi d\eta. \tag{8.31}
$$

EXAMPLE 8.5: Textbook Exercise 8.5

Prove that by assuming a constant pressure interpolation over the integration volume in Equation (8.11), a constant pressure technique equivalent to the mean dilatation method is obtained.

Solution

If p and δp are constant fields in textbook Equation (8.41) over the integration volume, then

$$
\int_V \left[(J-1) - \frac{p}{\kappa} \right] dV = 0, \tag{8.32}
$$

which gives

$$
\begin{aligned}
\frac{p}{\kappa} \int_V dV &= \int_V (J-1) dV \\
&= \int_V J dV - \int_V dV.
\end{aligned} \tag{8.33}
$$

Noting that $\int_V J dV = v$ and that $\int_V dV = V$ gives

$$\frac{p}{\kappa} V = v - V \quad \text{or} \quad p = \kappa \frac{v - V}{V}, \tag{8.34}$$

which corresponds to the mean dilatation technique expression for the pressure given in textbook Equation (8.12).

EXAMPLE 8.6: Textbook Exercise 8.6

A six-field Hu–Washizu type of variational principle with independent volumetric and deviatoric variables is given as

$$\begin{aligned}\Pi_{\text{HW}}(\boldsymbol{\phi}, \bar{J}, \boldsymbol{F}, p, \boldsymbol{P}', \gamma) &= \int_V \hat{\Psi}(\boldsymbol{C})\, dV + \int_V U(\bar{J})\, dV \\ &+ \int_V p(J - \bar{J})\, dV + \int_V \boldsymbol{P}' : (\boldsymbol{\nabla}_0 \boldsymbol{\phi} - \boldsymbol{F})\, dV \\ &+ \int_V \gamma \boldsymbol{P}' : \boldsymbol{F}\, dV - \Pi_{\text{ext}}(\boldsymbol{\phi}),\end{aligned}$$

where $\boldsymbol{C} = \boldsymbol{F}^{\text{T}} \boldsymbol{F}$, $J = \det(\boldsymbol{\nabla}_0 \boldsymbol{\phi})$, and $\boldsymbol{P}'$ denotes the deviatoric component of the first Piola–Kirchhoff stress tensor. Find the stationary conditions with respect to each variable. Explain the need to introduce the Lagrange multiplier γ. Derive the formulation that results from assuming that all the fields except for the motion are constant over the integration volume.[1]

Solution

The linearization with respect to $\boldsymbol{\phi}$ in the direction $\delta\boldsymbol{v}$ gives the Principle of Virtual Work in the usual fashion as

$$\begin{aligned}D\Pi_{\text{HW}}[\delta\boldsymbol{v}] &= \int_V \boldsymbol{P}' : \boldsymbol{\nabla}_0 \delta\boldsymbol{v}\, dV + \int_V p DJ[\delta\boldsymbol{v}]\, dV - D\Pi_{\text{ext}}[\delta\boldsymbol{v}] \\ &= \int_V \boldsymbol{P}' : \boldsymbol{\nabla}_0 \delta\boldsymbol{v}\, dV + \int_V p J \text{div}\, \delta\boldsymbol{v}\, dV - D\Pi_{\text{ext}}[\delta\boldsymbol{v}].\end{aligned} \tag{8.35}$$

Note, however, that the deviatoric and volumetric components of the internal energy are stated separately (see below Equations (8.40)).

[1] This exercise relates to J. Bonet and P. Bhargava, "A uniform deformation hexahedron element with hourglass control," *International Journal for Numerical Methods in Engineering* **36**(16), p. 2809 (1995).

Linearization with respect to $\boldsymbol{P}'$ and p now gives

$$D\Pi_{\mathrm{HW}}[\delta \boldsymbol{P}'] = \int_V \delta \boldsymbol{P}' : [\boldsymbol{\nabla}_0 \boldsymbol{\phi} - (1-\gamma)\boldsymbol{F}]\, dV = 0, \tag{8.36}$$

$$D\Pi_{\mathrm{HW}}[\delta p] = \int_V \delta p (J - \bar{J})\, dV = 0, \tag{8.37}$$

which establish relationships between $\boldsymbol{\nabla}_0 \boldsymbol{\phi}$ and $\boldsymbol{F}$ and between $J = \det[\boldsymbol{\nabla}_0 \boldsymbol{\phi}]$ and $\bar{J}$. For instance, taking $\boldsymbol{P}'$ and p constant over the integration volume as in textbook Section 8.6.5 gives

$$\bar{J} = \frac{1}{V}\int_V J\, dV = \frac{v}{V}, \tag{8.38}$$

$$\boldsymbol{F} = (1-\gamma)^{-1}\frac{1}{V}\int_V \boldsymbol{\nabla}_0 \boldsymbol{\phi}\, dV = (1-\gamma)^{-1}\overline{\boldsymbol{F}}. \tag{8.39}$$

The derivative in the direction of $\delta \boldsymbol{F}$ and $\delta \bar{J}$ gives the deviatoric and volumetric constitutive equations as

$$D\Pi_{\mathrm{HW}}[\delta \boldsymbol{F}] = \int_V \delta \boldsymbol{F} : \left[\frac{\partial \hat{\Psi}}{\partial \boldsymbol{F}} - (1-\gamma)\boldsymbol{P}'\right] dV = 0, \tag{8.40a}$$

$$D\Pi_{\mathrm{HW}}[\delta \bar{J}] = \int_V \delta \bar{J}\left(\frac{dv}{d\bar{J}} - p\right) dV = 0. \tag{8.40b}$$

For constant fields over the integration volume these yield

$$\boldsymbol{P}' = (1-\gamma)^{-1}\frac{\partial \hat{\Psi}}{\partial \boldsymbol{F}}\bigg|_{\boldsymbol{F}=(1-\gamma)^{-1}\overline{\boldsymbol{F}}}, \tag{8.41}$$

$$p = \frac{dv}{d\bar{J}}\bigg|_{\bar{J}=v/V}. \tag{8.42}$$

The final term leads to the condition

$$D\Pi_{\mathrm{HW}}[\delta\gamma] = \int_V \delta\gamma \boldsymbol{P}' : \boldsymbol{F}\, dV = 0, \tag{8.43}$$

which enforces the deviatoric condition on $\boldsymbol{P}'$ (see textbook Equation (8.3)). For constant fields this will lead to

$$\boldsymbol{P}' : \boldsymbol{F} = 0, \tag{8.44}$$

which, combined with the above Equation (8.41) for $\boldsymbol{P}'$, would enable the evaluation of γ. However, if $\hat{\Psi}$ is defined so that it is homogeneous of order 0, then its derivative is homogeneous of order -1 and the above equation for $\boldsymbol{P}'$ becomes

$$\boldsymbol{P}' = (1-\gamma)^{-1} \left.\frac{\partial \hat{\Psi}}{\partial \boldsymbol{F}}\right|_{\boldsymbol{F}=(1-\gamma)^{-1}\overline{\boldsymbol{F}}} = \left.\frac{\partial \hat{\Psi}}{\partial \boldsymbol{F}}\right|_{\overline{\boldsymbol{F}}}. \tag{8.45}$$

In other words, $\boldsymbol{P}'$ is independent of the volumetric component of $\boldsymbol{F}$. For consistency one can set $\boldsymbol{F}$ such that

$$\det \boldsymbol{F} = \bar{J}, \tag{8.46}$$

by setting

$$\boldsymbol{F} = \bar{J}^{1/3}[\det \overline{\boldsymbol{F}}]^{-1/3}\overline{\boldsymbol{F}}. \tag{8.47}$$

CHAPTER NINE

DISCRETIZATION AND SOLUTION

In this chapter the formulations presented in the previous chapter are finally realized in terms of a solution process based on the finite element method. Examples are given of the calculation of the deformation gradient, stresses, equivalent nodal forces, and tangent matrix components.

Equation summary

Spatial elasticity tensor [6.14]

$$\mathcal{c} = J^{-1}\phi_*[\mathcal{C}]$$

$$= \sum_{\substack{i,j,k,l=1\\ I,J,K,L=1}}^{3} J^{-1}F_{iI}F_{jJ}F_{kK}F_{lL}\,\mathcal{C}_{IJKL}\,\boldsymbol{e}_i \otimes \boldsymbol{e}_j \otimes \boldsymbol{e}_k \otimes \boldsymbol{e}_l. \tag{9.1}$$

Compressible neo-Hookean material second Piola–Kirchhoff stress tensor [6.28]

$$\boldsymbol{S} = \mu(\boldsymbol{I} - \boldsymbol{C}^{-1}) + \lambda(\ln J)\boldsymbol{C}^{-1}. \tag{9.2}$$

Compressible neo-Hookean material Cauchy stress tensor [6.29]

$$\boldsymbol{\sigma} = \frac{\mu}{J}(\boldsymbol{b} - \boldsymbol{I}) + \frac{\lambda}{J}(\ln J)\boldsymbol{I}. \tag{9.3}$$

Compressible neo-Hookean material elasticity tensor [6.40]

$$\mathcal{c}_{ijkl} = \lambda' \delta_{ij}\delta_{kl} + \mu' \left(\delta_{ik}\delta_{jl} + \delta_{il}\delta_{jk}\right), \tag{9.4}$$

where the effective coefficients λ' and μ' are [6.41]

$$\lambda' = \frac{\lambda}{J}; \quad \mu' = \frac{\mu - \lambda \ln J}{J}. \tag{9.5}$$

Spatial elasticity tensor symmetries [6.42]

$$\mathcal{c}_{ijkl} = \mathcal{c}_{klij} = \mathcal{c}_{jikl} = \mathcal{c}_{ijlk}. \tag{9.6}$$

Surface pressure component of the external virtual work [8.18]

$$\delta W^{p}_{\text{ext}}(\boldsymbol{\phi}, \delta \boldsymbol{v}) = \int_a p\boldsymbol{n} \cdot \delta \boldsymbol{v}\, da. \tag{9.7}$$

Deformation gradient tensor [9.5]

$$\boldsymbol{F} = \sum_{a=1}^{n} \boldsymbol{x}_a \otimes \boldsymbol{\nabla}_0 N_a. \tag{9.8}$$

Shape function and material coordinate derivatives [9.6a,b]

$$\frac{\partial N_a}{\partial \boldsymbol{X}} = \left(\frac{\partial \boldsymbol{X}}{\partial \boldsymbol{\xi}}\right)^{-T} \frac{\partial N_a}{\partial \boldsymbol{\xi}}\,; \quad \frac{\partial \boldsymbol{X}}{\partial \boldsymbol{\xi}} = \sum_{a=1}^{n} \boldsymbol{X}_a \otimes \boldsymbol{\nabla}_\xi N_a. \tag{9.9}$$

Equivalent nodal force [9.15b]

$$\boldsymbol{T}_a^{(e)} = \int_{v^{(e)}} \boldsymbol{\sigma} \boldsymbol{\nabla} N_a\, dv = \int_{v^{(e)}} \boldsymbol{\sigma} \frac{\partial N_a}{\partial \boldsymbol{x}}\, dv. \tag{9.10}$$

Constitutive component of the tangent stiffness matrix [9.35]

$$[\boldsymbol{K}_{c,ab}^{(e)}]_{ij} = \int_{v^{(e)}} \sum_{k,l=1}^{3} \frac{\partial N_a}{\partial x_k}\, \mathcal{c}_{ikjl}\, \frac{\partial N_b}{\partial x_l}\, dv\,; \qquad i,j = 1,2,3. \tag{9.11}$$

Initial stress component of the tangent stiffness matrix [9.44b]

$$\boldsymbol{K}_{\sigma,ab}^{(e)} = \int_{v^{(e)}} (\boldsymbol{\nabla} N_a \cdot \boldsymbol{\sigma} \boldsymbol{\nabla} N_b) \boldsymbol{I}\, dv. \tag{9.12}$$

EXAMPLE 9.1

A three-noded plane strain linear triangle finite element of unit thickness is deformed as shown in Figure 9.1. The material is defined by a compressible

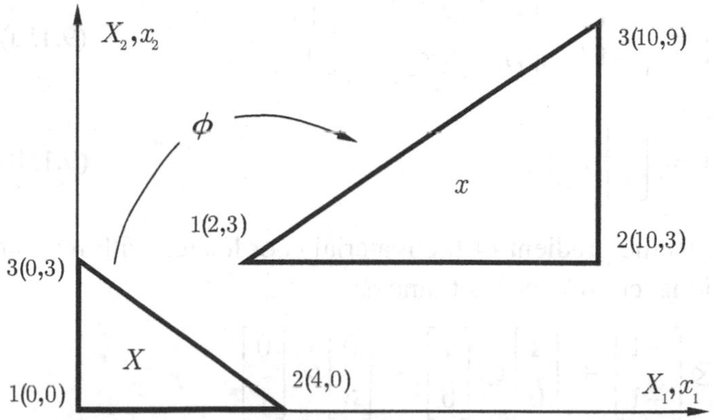

FIGURE 9.1 Three-node linear triangle

neo-Hookean material with $\lambda = 2$ and $\mu = 3$, see Equation (9.3). This example has the same geometry as textbook Example 9.1.

Calculate the the following items:

(a) deformation gradient tensor $\boldsymbol{F}$,
(b) Cauchy–Green tensors $\boldsymbol{C}$ and $\boldsymbol{b}$,
(c) second Piola–Kirchhoff and Cauchy stress tensors, $\boldsymbol{S}$ and $\boldsymbol{\sigma}$ respectively,
(d) vector of internal nodal, $\boldsymbol{T}^a$ forces for each node a,
(e) component of the tangent stiffness, $\boldsymbol{K}_{23}$ connecting nodes 2 – 3.

Solution

(a) The material and spatial coordinates, $\boldsymbol{X}_a$ and $\boldsymbol{x}_a$, $a = 1, 2, 3$ respectively, are listed as

$$\boldsymbol{X}_1 = \begin{bmatrix} 0 \\ 0 \end{bmatrix} ; \boldsymbol{X}_2 = \begin{bmatrix} 4 \\ 0 \end{bmatrix} ; \boldsymbol{X}_3 = \begin{bmatrix} 0 \\ 3 \end{bmatrix}, \tag{9.13a}$$

$$\boldsymbol{x}_1 = \begin{bmatrix} 2 \\ 3 \end{bmatrix} ; \boldsymbol{x}_2 = \begin{bmatrix} 10 \\ 3 \end{bmatrix} ; \boldsymbol{x}_3 = \begin{bmatrix} 10 \\ 9 \end{bmatrix}. \tag{9.13b}$$

The relevant shape functions and gradients with respect to the non-dimensional isoparametric coordinates are

$$N_1(\xi_1, \xi_2) = 1 - \xi_1 - \xi_2, \tag{9.14a}$$

$$N_2(\xi_1, \xi_2) = \xi_1, \tag{9.14b}$$

$$N_3(\xi_1, \xi_2) = \xi_2. \tag{9.14c}$$

$$\begin{bmatrix} \partial N_1/\partial \xi_1 \\ \partial N_1/\partial \xi_2 \end{bmatrix} = \begin{bmatrix} -1 \\ -1 \end{bmatrix}; \begin{bmatrix} \partial N_2/\partial \xi_1 \\ \partial N_2/\partial \xi_2 \end{bmatrix} = \begin{bmatrix} 1 \\ 0 \end{bmatrix}, \tag{9.15a}$$

$$\begin{bmatrix} \partial N_3/\partial \xi_1 \\ \partial N_3/\partial \xi_2 \end{bmatrix} = \begin{bmatrix} 0 \\ 1 \end{bmatrix}. \tag{9.15b}$$

From Equation (9.9) the gradient of the material coordinates with respect to the nondimensional coordinates is found as

$$\frac{\partial \boldsymbol{X}}{\partial \boldsymbol{\xi}} = \begin{bmatrix} 0 \\ 0 \end{bmatrix} \otimes \begin{bmatrix} -1 \\ -1 \end{bmatrix} + \begin{bmatrix} 4 \\ 0 \end{bmatrix} \otimes \begin{bmatrix} 1 \\ 0 \end{bmatrix} + \begin{bmatrix} 0 \\ 3 \end{bmatrix} \otimes \begin{bmatrix} 0 \\ 1 \end{bmatrix}$$

$$= \begin{bmatrix} 4 & 0 \\ 0 & 3 \end{bmatrix}. \tag{9.16}$$

Similarly, with respect to the spatial coordinates

$$\frac{\partial \boldsymbol{x}}{\partial \boldsymbol{\xi}} = \begin{bmatrix} 2 \\ 3 \end{bmatrix} \otimes \begin{bmatrix} -1 \\ -1 \end{bmatrix} + \begin{bmatrix} 10 \\ 3 \end{bmatrix} \otimes \begin{bmatrix} 1 \\ 0 \end{bmatrix} + \begin{bmatrix} 10 \\ 9 \end{bmatrix} \otimes \begin{bmatrix} 0 \\ 1 \end{bmatrix}$$

$$= \begin{bmatrix} 8 & 8 \\ 0 & 6 \end{bmatrix}. \tag{9.17}$$

Prior to calculating the derivatives of the shape functions with respect to the material and spatial coordinates, the transpose of the inverse of Equations (9.16) and (9.17) are simply

$$\left(\frac{\partial \boldsymbol{X}}{\partial \boldsymbol{\xi}}\right)^{-T} = \begin{bmatrix} \frac{1}{4} & 0 \\ 0 & \frac{1}{3} \end{bmatrix}; \left(\frac{\partial \boldsymbol{x}}{\partial \boldsymbol{\xi}}\right)^{-T} = \begin{bmatrix} \frac{1}{8} & 0 \\ -\frac{1}{6} & \frac{1}{6} \end{bmatrix}. \tag{9.18}$$

Using Equation (9.9), the gradients of the shape functions with respect to the material and spatial coordinates are now found, for example for node 1, as

$$\frac{\partial N_1}{\partial \boldsymbol{X}} = \left(\frac{\partial \boldsymbol{X}}{\partial \boldsymbol{\xi}}\right)^{-T} \frac{\partial N_1}{\partial \boldsymbol{\xi}} = \begin{bmatrix} \frac{1}{4} & 0 \\ 0 & \frac{1}{3} \end{bmatrix} \begin{bmatrix} -1 \\ -1 \end{bmatrix} = \begin{bmatrix} -\frac{1}{4} \\ -\frac{1}{3} \end{bmatrix}, \tag{9.19a}$$

$$\frac{\partial N_1}{\partial \boldsymbol{x}} = \left(\frac{\partial \boldsymbol{x}}{\partial \boldsymbol{\xi}}\right)^{-T} \frac{\partial N_1}{\partial \boldsymbol{\xi}} = \begin{bmatrix} \frac{1}{8} & 0 \\ -\frac{1}{6} & \frac{1}{6} \end{bmatrix} \begin{bmatrix} -1 \\ -1 \end{bmatrix} = \begin{bmatrix} -\frac{1}{8} \\ 0 \end{bmatrix}. \tag{9.19b}$$

The remaining gradients corresponding to Equation (9.19) for nodes 1 and 2 are

$$\frac{\partial N_2}{\partial \boldsymbol{X}} = \begin{bmatrix} \frac{1}{4} \\ 0 \end{bmatrix}; \quad \frac{\partial N_3}{\partial \boldsymbol{X}} = \begin{bmatrix} 0 \\ \frac{1}{3} \end{bmatrix}, \tag{9.20a}$$

$$\frac{\partial N_2}{\partial \boldsymbol{x}} = \begin{bmatrix} \frac{1}{8} \\ -\frac{1}{6} \end{bmatrix}; \quad \frac{\partial N_3}{\partial \boldsymbol{x}} = \begin{bmatrix} 0 \\ \frac{1}{6} \end{bmatrix}. \tag{9.20b}$$

The deformation gradient $\boldsymbol{F}$ can now be found using Equation (9.8) to yield

$$\boldsymbol{F} = \sum_{a=1}^{n} \boldsymbol{x}_a \otimes \boldsymbol{\nabla}_0 N_a \tag{9.21a}$$

$$= \begin{bmatrix} 2 \\ 3 \end{bmatrix} \otimes \begin{bmatrix} -\frac{1}{4} \\ -\frac{1}{3} \end{bmatrix} + \begin{bmatrix} 10 \\ 3 \end{bmatrix} \otimes \begin{bmatrix} \frac{1}{4} \\ 0 \end{bmatrix} + \begin{bmatrix} 10 \\ 9 \end{bmatrix} \otimes \begin{bmatrix} 0 \\ \frac{1}{3} \end{bmatrix}$$

$$= \begin{bmatrix} -\frac{1}{2} & -\frac{2}{3} \\ -\frac{3}{4} & -1 \end{bmatrix} + \begin{bmatrix} \frac{5}{2} & 0 \\ \frac{3}{4} & 0 \end{bmatrix} + \begin{bmatrix} 0 & \frac{10}{3} \\ 0 & 3 \end{bmatrix} \tag{9.21b}$$

$$= \begin{bmatrix} 2 & \frac{8}{3} \\ 0 & 2 \end{bmatrix}. \tag{9.21c}$$

(b) The right Cauchy–Green tensor is now easily found using Equation (4.4) as

$$\boldsymbol{C} = \boldsymbol{F}^{\mathrm{T}}\boldsymbol{F} = \begin{bmatrix} 2 & 0 \\ \frac{8}{3} & 2 \end{bmatrix} \begin{bmatrix} 2 & \frac{8}{3} \\ 0 & 2 \end{bmatrix} = \begin{bmatrix} 4 & \frac{16}{3} \\ \frac{16}{3} & \frac{100}{9} \end{bmatrix}, \tag{9.22}$$

and the left Cauchy–Green tensor, see Equation (4.39), is

$$\boldsymbol{b} = \boldsymbol{F}\boldsymbol{F}^{\mathrm{T}} = \begin{bmatrix} 2 & \frac{8}{3} \\ 0 & 2 \end{bmatrix} \begin{bmatrix} 2 & 0 \\ \frac{8}{3} & 2 \end{bmatrix} = \begin{bmatrix} \frac{100}{9} & \frac{16}{3} \\ \frac{16}{3} & 4 \end{bmatrix}. \tag{9.23}$$

(c) Recalling that the Jacobian $J = \det \boldsymbol{F} = 4$, $\lambda = 2$, and $\mu = 3$, the Cauchy stress tensor can be found from Equation (9.3) as

$$\boldsymbol{\sigma} = \frac{3}{4} \begin{bmatrix} \frac{100}{9} - 1 & \frac{16}{3} \\ \frac{16}{3} & 4 - 1 \end{bmatrix} + \frac{2}{4} \ln 4 \begin{bmatrix} 1 & 0 \\ 0 & 1 \end{bmatrix} = \begin{bmatrix} 8.2765 & 4 \\ 4 & 2.9431 \end{bmatrix}. \tag{9.24}$$

Using, Equation (9.2), the second Piola–Kirchhoff stress $\boldsymbol{S}$ is obtained as

$$\boldsymbol{S} = J\boldsymbol{F}^{-1}\boldsymbol{\sigma}\boldsymbol{F}^{\mathrm{T}} = \mu(\boldsymbol{I} - \boldsymbol{C}^{-1}) + \lambda(\ln J)\boldsymbol{C}^{-1}, \tag{9.25}$$

where

$$\boldsymbol{C}^{-1} = \boldsymbol{F}^{-1}\boldsymbol{F}^{-\mathrm{T}} = \begin{bmatrix} \frac{5}{36} & -\frac{1}{3} \\ -\frac{1}{3} & \frac{1}{4} \end{bmatrix}. \tag{9.26}$$

A simple substitution of Equation (9.26) in (9.25) yields

$$\begin{aligned} \boldsymbol{S} &= 3\begin{bmatrix} 1 - \frac{5}{36} & -\frac{1}{3} \\ -\frac{1}{3} & 1 - \frac{1}{4} \end{bmatrix} + 2\ln 4 \begin{bmatrix} \frac{5}{36} & -\frac{1}{3} \\ -\frac{1}{3} & \frac{1}{4} \end{bmatrix} \\ &= \begin{bmatrix} 2.8421 & 0.0758 \\ 0.0758 & 2.9431 \end{bmatrix}. \end{aligned} \tag{9.27}$$

(d) The equivalent nodal forces can be found using Equations (9.10), (9.24), and (9.19, 9.20). Conveniently for the linear triangular element, the Cauchy stress and shape function derivatives are constant over the element which, for example for node 1, enables the equivalent nodal force $\boldsymbol{T}_1^{(e)}$ at node 1 to be found as

$$\begin{aligned} \boldsymbol{T}_1^{(e)} &= \int_v \boldsymbol{\sigma}\frac{\partial N_1}{\partial \boldsymbol{x}}\,dv = 24\left(\boldsymbol{\sigma}\frac{\partial N_1}{\partial \boldsymbol{x}}\right) \\ &= 24\begin{bmatrix} 8.2765 & 4 \\ 4 & 2.9431 \end{bmatrix}\begin{bmatrix} -\frac{1}{8} \\ 0 \end{bmatrix} = \begin{bmatrix} -24.8294 \\ -12 \end{bmatrix}, \end{aligned} \tag{9.28}$$

where the spatial volume of the element is 24×1.

In a similar manner, the remaining equivalent nodal forces are calculated as

$$\boldsymbol{T}_2^{(e)} = \begin{bmatrix} 8.8294 \\ 0.2274 \end{bmatrix}; \quad \boldsymbol{T}_3^{(e)} = \begin{bmatrix} 16 \\ 11.7726 \end{bmatrix}. \tag{9.29}$$

Observe that the nodal forces are in equilibrium.

(e) The initial stress component of the tangent stiffness is the easier calculation and is obtained from Equation (9.12) for nodes 2 − 3 as

$$\boldsymbol{K}_{\sigma,23}^{(e)} = \int_v \left(\frac{\partial N_2}{\partial \boldsymbol{x}} \cdot \boldsymbol{\sigma} \frac{\partial N_3}{\partial \boldsymbol{x}} \right) \boldsymbol{I}\, dv = v^{(e)} \left(\frac{\partial N_2}{\partial \boldsymbol{x}} \cdot \boldsymbol{\sigma} \frac{\partial N_3}{\partial \boldsymbol{x}} \right) \boldsymbol{I} \tag{9.30a}$$

$$= 24 \left(\begin{bmatrix} \frac{1}{8}, & -\frac{1}{6} \end{bmatrix} \begin{bmatrix} 8.2765 & 4 \\ 4 & 2.9431 \end{bmatrix} \begin{bmatrix} 0 \\ \frac{1}{6} \end{bmatrix} \right) \begin{bmatrix} 1 & 0 \\ 0 & 1 \end{bmatrix}, \tag{9.30b}$$

$$= \begin{bmatrix} 0.0379 & 0 \\ 0 & 0.0379 \end{bmatrix}. \tag{9.30c}$$

The calculation of the (two-dimensional) constitutive component of the tangent stiffness matrix requires a temporary excursion into indicial notation. From Equations (9.11)

$$[\boldsymbol{K}_{c,23}]_{ij} = \int_{v^{(e)}} \sum_{k,l=1}^{2} \frac{\partial N_2}{\partial x_k} \, \mathcal{C}_{ikjl} \, \frac{\partial N_3}{\partial x_l} \, dv; \quad i,j = 1,2 \tag{9.31a}$$

$$= v^{(e)} \sum_{k,l=1}^{2} \left(\frac{\partial N_2}{\partial x_k} \frac{\partial N_3}{\partial x_l} \, \mathcal{C}_{ikjl} \right). \tag{9.31b}$$

After substituting Equation (9.4) into Equation (9.31) it is convenient to consider the λ' and μ' contributions within the summation separately to give

$$\sum_{k,l=1}^{2} \frac{\partial N_2}{\partial x_k} \frac{\partial N_3}{\partial x_l} \lambda' \delta_{ik} \delta_{jl} = \lambda' \frac{\partial N_2}{\partial x_i} \frac{\partial N_3}{\partial x_j}, \tag{9.32}$$

and

$$\sum_{k,l=1}^{2} \frac{\partial N_2}{\partial x_k} \frac{\partial N_3}{\partial x_l} \mu' \left(\delta_{ik} \delta_{jl} + \delta_{il} \delta_{jk} \right)$$

$$= \mu' \left[\left(\sum_{k,l=1}^{2} \frac{\partial N_2}{\partial x_k} \frac{\partial N_3}{\partial x_l} \right) \delta_{ij} + \frac{\partial N_3}{\partial x_i} \frac{\partial N_2}{\partial x_j} \right]. \tag{9.33}$$

Substituting Equations (9.32) and (9.33) into Equation (9.31) and reverting to direct notation yields

$$[\boldsymbol{K}_{c,23}] = \left[\lambda' \frac{\partial N_2}{\partial \boldsymbol{x}} \otimes \frac{\partial N_3}{\partial \boldsymbol{x}}\right] v^{(e)} + \mu' \left[\left(\frac{\partial N_2}{\partial \boldsymbol{x}} \cdot \frac{\partial N_3}{\partial \boldsymbol{x}}\right) \boldsymbol{I} + \frac{\partial N_3}{\partial \boldsymbol{x}} \otimes \frac{\partial N_2}{\partial \boldsymbol{x}}\right] v^{(e)}. \tag{9.34}$$

Substituting numerical data gives

$$[\boldsymbol{K}_{c,23}] = 24 \left[\frac{2}{4} \begin{bmatrix} \frac{1}{8} \\ -\frac{1}{6} \end{bmatrix} \otimes \begin{bmatrix} 0 \\ \frac{1}{6} \end{bmatrix}\right] + 24\left(\frac{(3 - 2\ln 4)}{4}\right)\left[\left(\begin{bmatrix} \frac{1}{8}, & -\frac{1}{6} \end{bmatrix} \begin{bmatrix} 0 \\ \frac{1}{6} \end{bmatrix}\right) \begin{bmatrix} 1 & 0 \\ 0 & 1 \end{bmatrix} + \begin{bmatrix} 0 \\ \frac{1}{6} \end{bmatrix} \otimes \begin{bmatrix} \frac{1}{8} \\ -\frac{1}{6} \end{bmatrix}\right] \tag{9.35}$$

$$= \begin{bmatrix} -0.0379 & 0.2500 \\ 0.0284 & -0.4091 \end{bmatrix}. \tag{9.36}$$

EXAMPLE 9.2: Textbook Exercise 9.1

Prove that the equivalent internal nodal forces can be expressed with respect to the initial configuration as

$$\boldsymbol{T}_a^{(e)} = \int_{V^{(e)}} \boldsymbol{F}\boldsymbol{S}\boldsymbol{\nabla}_0 N_a \, dV$$

and then confirm this equation by recalculating the equivalent nodal forces found in textbook Example 9.3.

Solution

From textbook Equation (9.10) the equivalent nodal force written with respect to the spatial coordinates is

$$\boldsymbol{T}_a^{(e)} = \int_{v^{(e)}} \boldsymbol{\sigma} \boldsymbol{\nabla} N_a \, dv \; ; \; \boldsymbol{\nabla} N_a = \left(\frac{\partial N_a}{\partial x_1}, \frac{\partial N_a}{\partial x_2}, \frac{\partial N_a}{\partial x_3}\right)^{\mathrm{T}}. \tag{9.37}$$

Recall that $dv = JdV$ and $\boldsymbol{\sigma} = J^{-1}\boldsymbol{FSF}^{\mathrm{T}}$ and note the components of $\boldsymbol{F}$ illustrated in textbook Example 4.2, page 100 (extended to 3×3), hence

$$\begin{aligned} \boldsymbol{T}_a^{(e)} &= \int_{V^{(e)}} \boldsymbol{FSF}^{\mathrm{T}} \boldsymbol{\nabla} N_a \, dV \\ &= \int_{V^{(e)}} \boldsymbol{FS} \, \boldsymbol{\nabla}_0 N_a \, dV. \end{aligned} \tag{9.38}$$

From textbook Examples 9.1 the material shape function derivatives deformation gradient (2×2) are

$$\frac{\partial N_1}{\partial \boldsymbol{X}} = -\frac{1}{12}\begin{bmatrix} 3 \\ 4 \end{bmatrix}; \quad \frac{\partial N_2}{\partial \boldsymbol{X}} = \frac{1}{12}\begin{bmatrix} 3 \\ 0 \end{bmatrix}; \quad \frac{\partial N_3}{\partial \boldsymbol{X}} = \frac{1}{12}\begin{bmatrix} 0 \\ 4 \end{bmatrix}, \tag{9.39a}$$

$$\boldsymbol{F} = \frac{1}{3}\begin{bmatrix} 6 & 8 \\ 0 & 6 \end{bmatrix}; \; J = \det \boldsymbol{F} = 4. \tag{9.39b}$$

To calculate the second Piola–Kirchhoff stress tensor $\boldsymbol{S}$, given in Equation (9.2) by

$$\boldsymbol{S} = \mu(\boldsymbol{I} - \boldsymbol{C}^{-1}) + \lambda(\ln J)\boldsymbol{C}^{-1}, \tag{9.40}$$

requires the determination of the right Cauchy–Green tensor $\boldsymbol{C}$, and its inverse by

$$\boldsymbol{C} = \boldsymbol{F}^{\mathrm{T}}\boldsymbol{F} = \frac{1}{9}\begin{bmatrix} 36 & 48 & 0 \\ 48 & 100 & 0 \\ 0 & 0 & 9 \end{bmatrix}, \tag{9.41a}$$

$$\boldsymbol{C}^{-1} = \boldsymbol{F}^{-1}\boldsymbol{F}^{-\mathrm{T}} = \frac{1}{36}\begin{bmatrix} 25 & -12 & 0 \\ -12 & 9 & 0 \\ 0 & 0 & 36 \end{bmatrix}, \tag{9.41b}$$

where the two-dimensional $\boldsymbol{C}$ has been extended to allow for plane strain behavior.[1]

For $\mu = 3$ and $\lambda = 2$, the second Piola–Kirchhoff stress tensor is

$$\boldsymbol{S} \approx \frac{1}{15}\begin{bmatrix} 40 & 0 & 0 \\ 0 & 45 & 0 \\ 0 & 0 & 48 \end{bmatrix}. \tag{9.42}$$

[1] Observe that $\boldsymbol{E} = \frac{1}{2}(\boldsymbol{C} - \boldsymbol{I})$; hence $\boldsymbol{E}_{33} = 0$.

For the initial configuration of the triangle shown in textbook Example 9.1, the volume is $6t$ where t is the thickness. Considering node 1, the equivalent nodal force is calculated to be

$$\begin{aligned} \boldsymbol{T}_1^{(e)} &= \int_{V^{(e)}} \boldsymbol{F}\boldsymbol{S}\,\boldsymbol{\nabla}_0 N_1\, dV \\ &= \boldsymbol{F}\boldsymbol{S}\,\boldsymbol{\nabla}_0 N_1\, V^{(e)}, \end{aligned} \tag{9.43}$$

which, considering only in, plane components of $\boldsymbol{S}$, yields

$$\begin{aligned} \boldsymbol{T}_1^{(e)} &= -\left(\frac{1}{540}\right)\begin{bmatrix} 6 & 8 \\ 0 & 6 \end{bmatrix}\begin{bmatrix} 40 & 0 \\ 0 & 45 \end{bmatrix}\begin{bmatrix} 3 \\ 4 \end{bmatrix} 6t \\ &= -\begin{bmatrix} 24t \\ 12t \end{bmatrix}. \end{aligned} \tag{9.44}$$

Similarly,

$$\boldsymbol{T}_2^{(e)} = \begin{bmatrix} 8t \\ 0t \end{bmatrix};\ \boldsymbol{T}_3^{(e)} \begin{bmatrix} 16t \\ 12t \end{bmatrix}. \tag{9.45}$$

EXAMPLE 9.3: Textbook Exercise 9.2

Prove that the initial stress matrix can be expressed with respect to the initial configuration as

$$\boldsymbol{K}_{\sigma,ab}^{(e)} = \int_{V^{(e)}} (\boldsymbol{\nabla}_0 N_a \cdot \boldsymbol{S}\boldsymbol{\nabla}_0 N_b)\boldsymbol{I}\, dV$$

and then confirm this equation by recalculating the initial stress matrix $\boldsymbol{K}_{\sigma,12}$ found in textbook Example 9.5.

Solution

From textbook Equation (9.12) the components of initial stress matrix with respect to the spatial configuration are

$$\boldsymbol{K}_{\sigma,ab}^{(e)} = \int_{v^{(e)}} (\boldsymbol{\nabla} N_a \cdot \boldsymbol{\sigma}\boldsymbol{\nabla} N_b)\boldsymbol{I}\, dv. \tag{9.46}$$

Recall that $dv = JdV$ and $\boldsymbol{\sigma} = J^{-1}\boldsymbol{F}\boldsymbol{S}\boldsymbol{F}^{\mathrm{T}}$. When substituted in the above equation, these give

$$\begin{aligned}
\boldsymbol{K}^{(e)}_{\sigma,ab} &= \int_{V^{(e)}} (\boldsymbol{\nabla} N_a \cdot (\boldsymbol{F}\boldsymbol{S}\boldsymbol{F}^{\mathrm{T}})\boldsymbol{\nabla} N_b)\boldsymbol{I}\, dV \\
&= \int_{V^{(e)}} ((\boldsymbol{F}^{\mathrm{T}}\boldsymbol{\nabla} N_a) \cdot \boldsymbol{S}(\boldsymbol{F}^{\mathrm{T}}\boldsymbol{\nabla} N_b))\boldsymbol{I}\, dV \\
&= \int_{V^{(e)}} (\boldsymbol{\nabla}_0 N_a \cdot \boldsymbol{S}\, \boldsymbol{\nabla}_0 N_b)\boldsymbol{I}\, dV.
\end{aligned} \tag{9.47}$$

For the particular case of the 1, 2 component,

$$\begin{aligned}
\boldsymbol{K}^{(e)}_{\sigma,12} &= \int_{V^{(e)}} (\boldsymbol{\nabla}_0 N_1 \cdot \boldsymbol{S}\, \boldsymbol{\nabla}_0 N_2)\boldsymbol{I}\, dV \\
&= (\boldsymbol{\nabla}_0 N_1 \cdot \boldsymbol{S}\, \boldsymbol{\nabla}_0 N_2)\boldsymbol{I}\, 6t \\
&= -\frac{1}{12}\begin{bmatrix} 3 & 4 \end{bmatrix}^{\mathrm{T}} \frac{1}{15}\begin{bmatrix} 40 & 0 \\ 0 & 45 \end{bmatrix} \frac{1}{12}\begin{bmatrix} 3 \\ 0 \end{bmatrix} \boldsymbol{I}\, 6t \\
&= \begin{bmatrix} -1t & 0 \\ 0 & -1t \end{bmatrix}.
\end{aligned} \tag{9.48}$$

EXAMPLE 9.4: Textbook Exercise 9.3

Show that the constitutive component of the tangent matrix can be expressed at the initial configuration by

$$\left[\boldsymbol{K}^{(e)}_{c,ab}\right]_{ij} = \sum_{I,J,K,L=1}^{3} \int_{V^{(e)}} F_{iI}\frac{\partial N_a}{\partial X_J}\mathcal{C}_{IJKL}\frac{\partial N_b}{\partial X_K}F_{jL}\, dV.$$

Solution

To conform to programming practice, this example will be re-couched to take advantage of the symmetries enshrined in the constitutive tensors. From Equation (9.11) the constitutive component of the tangent matrix relating node a to node b in element (e) in the spatial configuration becomes

$$\left[\boldsymbol{K}^{(e)}_{c,ab}\right]_{ij} = \int_{v^{(e)}} \frac{\partial N_a}{\partial x_k}\, \mathcal{c}_{ikjl}\, \frac{\partial N_b}{\partial x_l}\, dv, \tag{9.49}$$

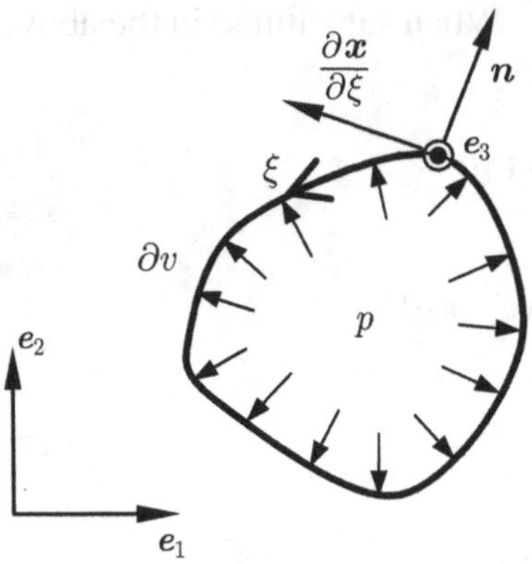

FIGURE 9.2 Two-dimensional pressure stiffness

where $i, j = 1, 2, 3$ and the symmetries in the spatial constitutive tensor $\mathcal{c}$ given by Equation (9.6) have been employed. From Equation (9.1) the spatial constitutive tensor $\mathcal{c}_{ikjl}$ is related to the corresponding Lagrangian form $\mathcal{C}_{IJKL}$ as

$$\mathcal{c}_{ijkl} = J^{-1}\phi_*[\mathcal{C}_{IJKL}] = J^{-1} F_{iI} F_{jJ} F_{kK} F_{lL}\, \mathcal{C}_{IJKL}. \tag{9.50}$$

Substituting Equation (9.50) into (9.49) and noting that $dv = JdV$ yields

$$\begin{aligned}\left[\boldsymbol{K}^{(e)}_{c,ab}\right]_{ij} &= \int_{V^{(e)}} \frac{\partial N_a}{\partial x_k} F_{iI} F_{jJ} F_{kK} F_{lL}\, \mathcal{C}_{IJKL} \frac{\partial N_b}{\partial x_l}\, dV \\ &= \int_{V^{(e)}} F_{iI} \frac{\partial N_a}{\partial X_k}\, \mathcal{C}_{IJKL} \frac{\partial N_b}{\partial X_l} F_{jJ}\, dV.\end{aligned} \tag{9.51}$$

EXAMPLE 9.5: Textbook Exercise 9.4

With the help of Example 8.2, derive a two-dimensional equation for the external pressure component of the tangent matrix $\boldsymbol{K}_p$ for a line element along an enclosed boundary of a two-dimensional body under uniform pressure p.

Solution

Recall from Equation (9.52) the virtual work expression relating to the traction force $p\boldsymbol{n}$ due to the uniform internal pressure p, see Figure 9.2, is

$$\delta W^p_{\text{ext}}(\phi, \delta \boldsymbol{v}) = \int_a p\boldsymbol{n} \cdot \delta \boldsymbol{v}\, da. \tag{9.52}$$

where $\boldsymbol{n} = (n_1, n_2)^{\mathrm{T}}$ and $\delta\boldsymbol{v} = (\delta v_1, \delta v_2)^{\mathrm{T}}$. Adopting a similar approach to that given in textbook Section 8.5.2, consider the boundary to be an isoparametric line element. For the two-dimensional situation with unit thickness, da becomes dl where dl is an elemental length on the parameterized boundary ∂v (see Equation (8.5) and Figure 8.1); consequently

$$\boldsymbol{n} = \frac{\frac{\partial \boldsymbol{x}}{\partial \xi} \times \boldsymbol{e}_3}{\left\| \frac{\partial \boldsymbol{x}}{\partial \xi} \times \boldsymbol{e}_3 \right\|}, \quad dl = \left\| \frac{\partial \boldsymbol{x}}{\partial \xi} \times \boldsymbol{e}_3 \right\| d\xi, \tag{9.53}$$

where $\partial \boldsymbol{x}/\partial \xi$ is the local tangent vector and $\boldsymbol{e}_3$ the unit base vector outward normal to the plane. The virtual work expression can now be written as

$$\delta W_{\text{ext}}^{p}(\boldsymbol{\phi}, \delta\boldsymbol{v}) = \int_{\partial v_\xi} p\, \delta\boldsymbol{v} \cdot \left(\frac{\partial \boldsymbol{x}}{\partial \xi} \times \boldsymbol{e}_3 \right) d\xi. \tag{9.54}$$

Linearization of the above expression yields

$$\begin{aligned} D\delta W_{\text{ext}}^{p}(\boldsymbol{\phi}, \delta\boldsymbol{v})[\boldsymbol{u}] &= D\left(\int_{\partial v_\xi} p\, \delta\boldsymbol{v} \cdot \left(\frac{\partial \boldsymbol{x}}{\partial \xi} \times \boldsymbol{e}_3 \right) d\xi \right) [\boldsymbol{u}] \\ &= \int_{\partial v_\xi} p\, \delta\boldsymbol{v} \cdot \left(\frac{\partial \boldsymbol{u}}{\partial \xi} \times \boldsymbol{e}_3 \right) d\xi \\ &= - \int_{\partial v_\xi} p\, \delta\boldsymbol{v} \cdot \left(\boldsymbol{e}_3 \times \frac{\partial \boldsymbol{u}}{\partial \xi} \right) d\xi \\ &= \int_{\partial v_\xi} p\, \boldsymbol{e}_3 \cdot \left(\delta\boldsymbol{v} \times \frac{\partial \boldsymbol{u}}{\partial \xi} \right) d\xi, \end{aligned} \tag{9.55}$$

where cyclic permutation has been applied. Observe that the last expression in the above equation is, in general, nonsymmetric as the exchange of $\delta\boldsymbol{v}$ and $\boldsymbol{u}$ does not result in the same outcome. The term $D\delta W_{\text{ext}}^{p}(\boldsymbol{\phi}, \delta\boldsymbol{v})[\boldsymbol{u}]$ gives the change in the virtual work due to a change in the spatial configuration given by $\boldsymbol{u}$ with the pressure being constant. After discretization this linearization gives the change in the equilibrating forces under the same conditions.

In order to incorporate the requirement that the boundary is enclosed, observe that

$$\int_{\partial v_\xi} \frac{\partial}{\partial \xi} (\delta\boldsymbol{v} \times \boldsymbol{u}) d\xi = 0. \tag{9.56}$$

Applying the scalar product of $p\,\boldsymbol{e}_3$ (which is not a function of ξ) to the above expression and using the chain rule gives

$$\begin{aligned}\int_{\partial v_\xi} p\,\boldsymbol{e}_3 \cdot \frac{\partial}{\partial \xi}(\delta\boldsymbol{v}\times\boldsymbol{u})d\xi &= \int_{\partial v_\xi} p\,\boldsymbol{e}_3 \cdot \left(\frac{\partial \delta\boldsymbol{v}}{\partial \xi}\times\boldsymbol{u}\right)d\xi \\ &\quad + \int_{\partial v_\xi} p\,\boldsymbol{e}_3 \cdot \left(\delta\boldsymbol{v}\times\frac{\partial \boldsymbol{u}}{\partial \xi}\right)d\xi \\ &= 0. \end{aligned} \tag{9.57}$$

Hence the linearization given in Equation (9.55) can be rewritten as

$$\int_{\partial v_\xi} p\,\boldsymbol{e}_3 \cdot \left(\delta\boldsymbol{v}\times\frac{\partial \boldsymbol{u}}{\partial \xi}\right)d\xi = -\int_{\partial v_\xi} p\,\boldsymbol{e}_3 \cdot \left(\frac{\partial \delta\boldsymbol{v}}{\partial \xi}\times\boldsymbol{u}\right)d\xi. \tag{9.58}$$

A symmetric expression for the linearization can now be developed by adding half of the left side of Equation (9.58) to half of the right side of Equation (9.58) to give

$$\begin{aligned} D\delta W^p_{\text{ext}}(\boldsymbol{\phi},\delta\boldsymbol{v})[\boldsymbol{u}] &= \frac{1}{2}\left(\int_{\partial v_\xi} p\,\boldsymbol{e}_3 \cdot \left(\delta\boldsymbol{v}\times\frac{\partial \boldsymbol{u}}{\partial \xi}\right)d\xi\right) \\ &\quad -\frac{1}{2}\left(\int_{\partial v_\xi} p\,\boldsymbol{e}_3 \cdot \left(\frac{\partial \delta\boldsymbol{v}}{\partial \xi}\times\boldsymbol{u}\right)d\xi\right) \\ &= \frac{1}{2}\left(\int_{\partial v_\xi} p\,\boldsymbol{e}_3 \cdot \left(\delta\boldsymbol{v}\times\frac{\partial \boldsymbol{u}}{\partial \xi}\right)d\xi\right) \\ &\quad +\frac{1}{2}\left(\int_{\partial v_\xi} p\,\boldsymbol{e}_3 \cdot \left(\boldsymbol{u}\times\frac{\partial \delta\boldsymbol{v}}{\partial \xi}\right)d\xi\right). \end{aligned} \tag{9.59}$$

Observe that in the above expression the interchange of $\delta\boldsymbol{v}$ and $\boldsymbol{u}$ in the second equation leaves the expression unchanged, indicating that discretization will yield a symmetric stiffness matrix. Written explicitly, the virtual work term is

$$D\delta W_{\mathrm{ext}}^{p}(\phi, \delta v)[u] = \frac{1}{2}\int_{\partial v_\xi} p\left(\delta v_1 \frac{\partial u_2}{\partial \xi} - \delta v_2 \frac{\partial u_1}{\partial \xi}\right) d\xi$$

$$+\frac{1}{2}\int_{\partial v_\xi} p\left(u_1 \frac{\partial \delta v_2}{\partial \xi} - u_2 \frac{\partial \delta v_1}{\partial \xi}\right) d\xi. \tag{9.60}$$

Applying spatial discretization yields

$$\begin{aligned} &D\delta W_{\mathrm{ext}}^{p}(\phi N_a \delta v_a)[N_b u_b] \\ &= \frac{1}{2}\int_{\partial v_\xi} p\left(\delta v_1^a N_a \frac{\partial N_b}{\partial \xi} u_2^b - \delta v_2^a N_a \frac{\partial N_b}{\partial \xi} u_1^b\right) d\xi \\ &\quad + \frac{1}{2}\int_{\partial v_\xi} p\left(\delta v_2^a N_a \frac{\partial N_b}{\partial \xi} u_1^b - \delta v_1^a N_a \frac{\partial N_b}{\partial \xi} u_2^b\right) d\xi \\ &= [\delta v_1^a, \delta v_2^a]\left[\boldsymbol{K}_{p,ab}^{(e)}\right]_{ij} \begin{bmatrix} u_1^b \\ u_2^b \end{bmatrix}, \end{aligned} \tag{9.61}$$

where

$$\left[\boldsymbol{K}_{p,ab}^{(e)}\right]_{ij} = \begin{bmatrix} 0 & \frac{1}{2}\left(N_a \frac{\partial N_b}{\partial \xi} - N_b \frac{\partial N_a}{\partial \xi}\right) \\ \frac{1}{2}\left(N_b \frac{\partial N_a}{\partial \xi} - N_a \frac{\partial N_b}{\partial \xi}\right) & 0 \end{bmatrix}. \tag{9.62}$$

Interchanging nodes a and b reveals that

$$\left[\boldsymbol{K}_{p,ab}^{(e)}\right]_{ij} = \left[\boldsymbol{K}_{p,ba}^{(e)}\right]_{ij}^{\mathrm{T}}. \tag{9.63}$$

EXAMPLE 9.6: Textbook Exercise 9.5

Recalling the line search method discussed in textbook Section 9.6.2, show that minimizing $\Pi(\eta) = \Pi(\mathbf{x}_k + \eta\mathbf{u})$ with respect to η gives the orthogonality condition

$$R(\eta) = \mathbf{u}^{\mathrm{T}}\mathbf{R}(\mathbf{x}_k + \eta\mathbf{u}) = 0.$$

Solution

Define a total potential energy function Π in terms of a single parameter η as follows:

$$\Pi(\eta) = \Pi(\mathbf{x}_k + \eta \mathbf{u}). \tag{9.64}$$

In order to obtain the minimum of the function $\Pi(\eta)$ at iterative position $\mathbf{x}_k$ in the given direction $\boldsymbol{u}$ (the current iterative change in $\mathbf{x}$), it must satisfy

$$\frac{d\Pi}{d\eta} = 0. \tag{9.65}$$

Applying the chain rule yields

$$\begin{aligned}\frac{d\Pi}{d\eta} &= \left(\left.\frac{\partial \Pi}{\partial \mathbf{x}}\right|_{\mathbf{x}_k + \eta \mathbf{u}} \right)^{\mathrm{T}} \frac{d\mathbf{x}}{d\eta}; \qquad \mathbf{x} = \mathbf{x}_k + \eta \mathbf{u} \\ &= \left.\frac{\partial \Pi}{\partial \mathbf{x}}\right|_{\mathbf{x}_k + \eta \mathbf{u}} \cdot \mathbf{u} \\ &= 0.\end{aligned} \tag{9.66}$$

Recalling the discussion in textbook Section 8.6.1, observe that the partial derivative in the above equation is equivalent to $D\Pi(\mathbf{x})[\mathbf{x}_k + \eta \mathbf{u}]$ which equals the residual force $\mathbf{R}(\mathbf{x}_k + \eta \mathbf{u})$. Consequently, minimizing $\Pi(\eta)$ gives

$$R(\eta) = \mathbf{u}^{\mathrm{T}} \mathbf{R}(\mathbf{x}_k + \eta \mathbf{u}) = 0. \tag{9.67}$$

Printed in the United States
by Baker & Taylor Publisher Services